# TINGED WITH GOLD

# TINGED WITH GOLD

❧ Hop Culture in the United States ❧

## MICHAEL A. TOMLAN

The University of Georgia Press

ATHENS AND LONDON

Designed by Kathi L. Dailey
Set in Century Expanded by Keystone Typesetters, Inc.
Printed and bound by Thomson-Shore, Inc.
The paper in this book meets the guidelines for
permanence and durability of the Committee on
Production Guidelines for Book Longevity of the
Council on Library Resources.

Printed in the United States of America

96  95  94  93  92    C    5  4  3  2  1

Library of Congress Cataloging in Publication Data

Tomlan, Michael A.
    Tinged with gold : hop culture in the United States / Michael
A. Tomlan.
      p.    cm.
    Includes bibliographical references and index.
    ISBN 0-8203-1313-0 (alk. paper)
    1. Hops—United States.   2. Hops—Social aspects—United
States.   3. Hop pickers—United States.   I. Title.
SB317.H64T65 1992
338.1'7382—dc20                                          90-46389
                                                              CIP

British Library Cataloging in Publication Data available

*Those were the days when the "hop was king," and the whole
countryside was one great hop yard, and beautiful. It was
the hop that built many of the big farm houses, now abandoned. Many a farmer made the value of his farm out of a
single good year's crop. When the time came for harvesting
the crop, the air of the town became tense; the housewives became worried as all the help insisted on a week off to go "hop
pickin'." There were rumors of great camps of tramps in the
woods about to raid the town; the police force of two men, one
with one arm and the other with one leg, became worried and
patrolled the town until one a.m. instead of the quitting at
eleven p.m., as was customary. No one went abroad after
dark unless armed with a pistol, more dangerous to the
owner than to his enemy. They were the halcyon days for the
boys and young men who tip-toed about the town looking for
the invaders and listening to the tales of the police patrol,
one with a club and one with a lantern. The street lamps
burned until twelve instead of being turned out at eleven,
and the whole atmosphere was one of suppressed alarm and
excitement. Thousands of tough "pickers" came from the
cities to earn the eighty cents a box for "pickin'," and to enjoy
the nightly barn dances given for their amusement.*

—James Fenimore Cooper, *Reminiscences of
Mid-Victorian Cooperstown*

# CONTENTS

# ILLUSTRATIONS

# ACKNOWLEDGMENTS

Over the course of the last several years a number of people have shared their information and insights with me. Although it is impossible to recognize them all, it is appropriate to mention those who have been particularly helpful.

Barbara Giambastiani, formerly the executive director of the Madison County Historical Society, not only was responsible for sponsoring a limited study that sparked the much longer investigation now finished, but also provided access to material she had collected. Harriet R. Rogers, formerly the president of the Town of Middlefield Historical Association in Otsego County, was especially generous with her time. Ms. Rogers made a tremendous effort to contact all the hop house owners in her area of Otsego County and accompanied the author in the field, literally and figuratively helping to open doors. John Alden Haight, professor emeritus of horticulture at the State University of New York, Morrisville, enthusiastically shared his knowledge of the upstate hop belt. His fervor on the topic of hop culture will not be forgotten. Sydney Erickson, president of the Waterville Historical Society, was helpful in identifying sites in Oneida County and in referring the author to other knowledgeable individuals.

This study has also benefited from the cooperation the author has received from a number of historians living in hop-growing regions outside the state of New York. Kirk Mohney, who wrote a fine master's thesis on hop culture in southern Oneida County, has since become an architectural historian for the Maine Historical Preservation Commission. He was helpful by ferreting out material in his corner of the country. Robert McCullough, a close observer of the Vermont landscape, went out of his way to provide information about material in that state. Lawrence Garfield, a staff member of the State Historical Society of Wisconsin in Madison, was cooperative by sending along references, as was Nijole Etzwiler of the Sauk County Historical Society in Baraboo, Wisconsin. Anyone doing research in Ukiah, California, inevitably comes to know Lila J. Lee, the energetic director of the Mendocino County Historical Society, who was very helpful to a stranger dropping in unannounced on an extremely hot day. Maura Johnson, from her desk in Eugene, Oregon, was ever willing to field check an example, find a footnote reference, or contribute a photograph, and she kept a sharp lookout for picture postcards of hop culture in her area. Frank Green, director of the Washington State Historical Society, was extremely cooperative in opening up the superb photographic collections in his charge. In England, Robert F. Farrar, curator of the Agricultural Museum, Wye College, Wye, near Ashford, Kent, was an excellent host and shared with me his observations from personal familiarity with hop culture. The anonymous scholars who read the manuscript for the University of Georgia Press provided insightful comments and were helpful by challenging me to sharpen several important points. Lastly, three professionals took on the challenge of production: Mary Raddant Tomlan created an excellent index; Kathi Dailey designed a book that maximized the potential of every illustration; and Madelaine Cooke, as managing editor, guided along the text with the right balance of patience and perseverance. It has been a pleasure to work with her.

To each of these persons, and the dozens of property owners who have patiently answered my questions and allowed me access to their property and buildings, I extend my heartfelt thanks. Even with all this help, there are undoubtedly errors and omissions that should be corrected, and these are my own.

ACKNOWLEDGMENTS

# ❧ Introduction ❦

Preceding page: *The hop
plant, a leafy climbing
vine, produces clusters of
cones at harvest time.
(Drawing by Belle
Hodgson, c. 1890.)*

Although scholars continue to examine many aspects of American history at length, the story of agricultural development in this country receives comparatively little attention. Scores of books contain material on political events, military battles, the lives of influential individuals, and the activities of urban ethnic groups, but the agrarian economy that was once everywhere evident goes largely unnoticed and unappreciated. There are at least two reasons for this. First, our urban and suburban society has limited contact with the decreasing number of people who turn the soil or raise livestock. Second, many once-common rural activities are no longer useful and have been discontinued. The average American drives through or flies over rural landscapes but has only a vague understanding of the activities that are carried on in them.

The structures that remain as vestigial reminders of agricultural development are likewise not widely understood. The history of architecture has an intrinsic urban bias, supported by the archival collections of community-centered historical organizations and the public record. The homes of the rich, commercial palaces, civic buildings, and churches are places of continuing interest and curiosity. Recently, industrial buildings, public works, and suburban domestic structures have been recognized as appropriate objects of study. However, rural agricultural buildings have received little attention, despite the recognition that they are fast disappearing.[1] Although their forms may be appreciated, these structures have few ornamental details to catch and hold the eye, and few documents to sustain an investigation.

Cultural geographers were the first to study rural vernacular buildings in this country as a legitimate academic endeavor, and their methodology shaped subsequent research. In 1925 Carl Sauer focused attention on vernacular housing as an important three-dimensional record in the landscape.[2] He believed that rural buildings reflected the organization of families, expressed the conditions of the local economy, and embodied traditions that were not evident otherwise. Fred Kniffen studied folk structures throughout the eastern United States from the colonial period to the late nineteenth century. In an article published in 1965, Kniffen demonstrated that by sorting houses and barns into types and plotting them on maps, he could compare their distribution with maps based on dialect and other cultural patterns. In this manner he tracked the diffusion of ideas about building form from various "cultural hearths" to newly settled regions. Kniffen's ideas about the diffusion and morphological development of crib barns remain one of the most striking contributions to the understanding of vernacular architecture.[3] A few years later, Henry Glassie, acknowledging his debt to Kniffen, likewise attempted to discover the patterns in material culture in the Northeast.[4] One of Glassie's suggestions was that one could find patterns within regions by studying the material culture of specialty crops. He mentioned tobacco as one example and noted the common appearance of flue-cured tobacco barns in several areas of the Southeast. Further, he indicated that "parts of Central New York can be cut out of the Northeast and drawn together on the basis of hops."[5] His subsequent work, however, focused on the architectural "grammar" he found in the geometry of vernacular buildings.[6]

Although a number of other geographers have dealt with the economic or social aspects of hop culture in various areas of the country,[7] only two have attempted to deal with hop structures. Charles Calkins and William Laatch followed the time-honored approach of field-surveying Waukesha County, Wisconsin, in search of hop kilns.[8] On the basis of the remaining physical evidence, they concluded that the hop structures in that area were comparatively small buildings that owed little to the structures in the East but instead were directly inspired by those in England. Herein lies a problem of methodology. Is the information about the handful of remaining structures and limited number of archeological sites sufficient to justify this conclusion? The historical evidence offered in the work at hand demonstrates the close working relationship between Wisconsin growers and those in the eastern United States as well as an awareness of English activity. Hence it appears that Calkins and Laatch were misled by relying too heavily on the available physical evidence.

More recently, another geographer, Allen Noble, assembled a considerable amount of the information about the North American farm barn.[9] Acknowledging the need for more study, he examined the diffusion of several barn types. When mentioning hop houses, Noble indicated that the crop was grown in both the eastern and western United States, and he provided drawings of several previously published images. However, he

TINGED WITH GOLD

avoided any discussion of the diffusion of models. The disparate forms he found defied easy summary.

Although the picture painted by cultural geographers contains a useful perspective, architectural historians interested in construction offered other approaches. The search for cultural patterns in the rural landscape led these researchers to investigate similarities that derived from the settlers' common ethnic background. English, German, and Dutch barns were recorded, mapped, and analyzed. For example, the architects and architectural historians Theodore H. M. Prudon and John Fitchen studied Dutch barns, primarily for their structural systems.[10] Framing, joinery, and the method of construction are the elements displayed in common by these eighteenth-century barns found in the Mohawk Valley and northern New Jersey. The time period and the geographic distribution of the buildings set logical limitations in these studies.

Within the last decade historians of vernacular architecture have shown that if their subject is to be fully understood, a more thorough historical investigation of the context is needed. The methodology adopted in the present work recognizes the importance of providing a broad overview. Although the diffusion of ideas from one place to another is a central theme, census results rather than field survey results have been relied on to determine where the crop was grown. This gives a more accurate indication of the original distribution of the buildings used to collect, dry, bale, and store hops. Further, by comparing the results across time, the rise and fall of a number of production centers can be followed. Hence, economic history permits the investigator to fit any surviving artifact into a spatial and chronological framework, making its local history more meaningful.

The intent of this study is not only to demonstrate a more comprehensive methodology for studying agrarian building types, but also to advance an awareness and understanding of hop culture and to document and interpret the structures and artifacts that remain.

Hop culture is important in the history of agriculture because hops were among the first specialty crops to attract widespread interest among enterprising, progressive farmers. Hop growing required an unusually sophisticated understanding of plant science, drying technology, and market economics. Hops are known to have been indigenous to parts of Asia,

Europe, and North America. These small cones, growing in clusters along leafy, climbing vines, were sometimes collected for medicinal purposes, but by the time the English colonists came to the New World, hops were valued chiefly as an additive in beer making. In the early nineteenth century, as what was once a modest home activity grew into the beer-making industry, the demand for hops increased dramatically. Hop culture had a profound economic impact on all the regions in which it was cultivated. It brought some growers unheard of wealth almost overnight, while other hop farmers and dealers slid into ruinous debt just as quickly. Meanwhile, hoards of pickers would patronize local shops and stores, providing the equivalent of a Christmas shopping season at the end of the summer.

As defined here, hop culture includes not only the growth, cultivation, and harvesting of the plant, but also the economic, social, and recreational activities of the people who became involved in the various processes and procedures dealing with the crop. It includes a record of the mechanical inventions, the technical developments, and the architectural traditions that shaped hop kilns, hop houses, and hop dryers and coolers in several states.

To understand these buildings properly it is necessary to reconstruct the context in which they were built. The first chapter explores the reasons for the establishment, growth, and precipitous decline of the hop industry in the northeastern United States. An examination of its subsequent rise and gradual fall in the West follows. This review is intended to acquaint the reader with the principal districts, counties, townships, villages, and cities that shared a set of concerns about this crop. It introduces the pioneers in each of the hop-growing regions, tracing their roots to explore how ideas spread from one area to another.

The story has discretely defined geographical and chronological limits. Though planted by the first colonists, hops were produced commercially in New England only after the Revolutionary War. By the mid-1840s New York had become the nation's leading producer of hops, a position it occupied unrivaled for over fifty years. Michigan and Wisconsin briefly caught the "fever," but California, Oregon, and Washington led the way in the present century, and there the memories of hop culture are clearest. Indeed, it is only in the Pacific Northwest that this crop remains economically viable.

The second chapter is devoted to reviewing the steps involved in growing and harvesting. Hops require special attention in siting, laying out the field, planting, training and cultivating, poling or stringing, picking, curing, and bailing the crops. All of these activities led to experimentation and technological developments that are evident in the landscape.

The people involved in the process were no less important. Two groups were of particular importance: the growers and the pickers. A chapter devoted to the growers focuses on such concerns as the purposes in planting the crop, the economics of hop growing and marketing, and the involvement of agents and middlemen in promoting the culture. Another chapter examines the attractions and disadvantages of picking from the point of view of those who were in the fields, the ethnicity of the people involved, and the tensions that built up in what was sometimes billed as a time of recreation and reward. All of these aspects of hop culture varied considerably from one era on one side of the country, to another period some three thousand miles distant.

Last, the focus shifts to the manner in which the curing and baling processes were accommodated in the design of buildings used for drying and storing hops. An examination of contemporary agricultural literature and an investigation of a number of examples that were built provides insight and understanding of the form, dimensions, plan, arrangement, structural configuration, and materials used in hop buildings. This, in turn, enables ideas to be traced and a number of comparisons to be made. Further, the function of the areas or spaces around the structures can be shown to be as important as the activities that went on within the buildings, because these external relationships dictated the position of other structures and objects in the vicinity.

This broad interdisciplinary study should make it easier to identify these special agrarian buildings and their sites. It is hoped that the reader will also gain an understanding of the importance of the buildings. This understanding will encourage not only further study and appreciation of hop buildings but also their preservation.

# ❧1❧

# The History of Hop Growing
# in the United States

The history of hop growing in what would become the United States began shortly after the first European settlers landed. Beer making was common in most of the countries of northern Europe, and brewing became a household occupation in many of the colonies along the East Coast.[1] Because beer was in demand, hops were considered a valuable asset, whether grown domestically in the kitchen garden, discovered wild in the woods, or imported.[2]

The Dutch were probably the first to develop a significant interest in brewing on the North American continent.[3] After landing on Manhattan Island in 1607, they took only two decades to establish a relatively solid agricultural base, harvesting quantities of grain. A brewery was erected in 1633 on the plantation of the director general of New Amsterdam, near the fort. A few years later, taxes were levied on the production of beer and wine; furthermore, brewers were forbidden to retail beer and tavern keepers could not manufacture it.[4]

Travelers noted that native hops were readily available. In 1642, for example, David Pietersz De Vries admired the wheat, rye, barley, oats, and peas that his countrymen were growing. In addition he noted that "our Netherlanders . . . can brew as good beer here as in our own Fatherland, for hops grow in the woods."[5] However, the Dutch settlers never capitalized on this piece of good fortune. On the contrary, hops were shipped regularly from the mother country. Dutch colonists who settled on the Delaware River, for example, were required by their sponsor, the city of Amsterdam, to import any hops that were needed. Imported hops were more expensive than the locally produced article; by 1646 the prices for both were regularly posted at markets in Manhattan.[6] Apparently the Dutch control over the sale of hops and the production of malt beverages effectively foreclosed the possibility of any commercial development.

The colonists in New England were allowed to be more enterprising. As early as 1628 the Endicott expedition introduced hops to Massachusetts Bay.[7] Although penalties for drunkenness were severe, Massachusetts promoted beer making in the belief that malt liquors were healthy and provided the best means to wean people away from the dangers of ardent spirits. Unlike most of the other colonies, Massachusetts even permitted innkeepers to produce beer. As a result, the breweries of

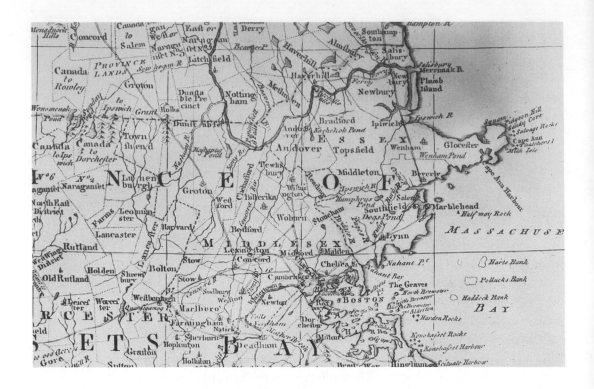

*1.1 Middlesex County, Massachusetts, and the surrounding area at the time of the Revolution. Wilmington, Tewkesbury, and Townsend were the leading market towns in the hop-growing region of the state. (From* Bowles' New Pocket Map of New England, *c. 1775. Courtesy of John W. Reps.)*

the Boston area not only imported hops but also purchased them from progressive farmers in nearby towns.

Growing grain in New England during the seventeenth century was not easy. The extremes in weather, which was either too hot or too cold, too wet or too dry, the insect damage, and plant disease were often thought to be the will of God. However, modest amounts of surplus grain were exported, as were hops. By 1718, for example, a Massachusetts cargo ship bound for Newfoundland included hops. In 1763 the schooner *Bernard* carried three thousand pounds of hops from Massachusetts to New York. Growers knew that breweries to the south, particularly those in Philadelphia, were anxious to obtain the finest cones available.[8]

From the middle of the eighteenth century until the early nineteenth century, Massachusetts was the acknowledged leader in hop production in North America. Middlesex County in particular was famous for its hop yards, and Wilmington was the first place in which the culture grew to a fever pitch.[9] By today's standards, the level of production was rather modest; in late-eighteenth-century terms, however, the output was already very impressive. In 1780, at a meeting of the eight or ten major growers in the town, every producer was listed and the quantity each produced was given. In that year they harvested a total of about thirty thousand pounds.[10]

The culture soon spread as farmers from neighboring towns went to Wilmington, known as "Hop Town," for their roots. The comparatively short distance between Wilmington and Boston, only sixteen miles, coupled with access to the Middlesex Canal, guaranteed easy transportation to both domestic and foreign markets. Tewkesbury, about twenty-four miles northwest of Boston; Shirley, some thirty-eight miles out in the same direction; and Townsend, still farther away, also became important centers. The generally level terrain and well-drained, light soil favored all four towns, which were known for raising fine field crops.[11]

By the end of the eighteenth century, a considerable portion of the Massachusetts crop was being exported to France and Germany. To improve market conditions, in 1806 the commonwealth mandated the compulsory inspection and grading of all hops. A high standard was established, and the character of official inspection so impressed European buyers that they regularly requested the "first sort" Massachusetts variety. This, in turn, allowed the sellers to charge a higher price.[12]

For the next three decades, Massachusetts continued to export hops to breweries in Europe and along the East Coast.[13] Several English crop failures in the early nineteenth century provided the impetus for farmers throughout New England to try their hand. Glowing reports of enormous profits were common, and production spread beyond Massachusetts to New Hampshire, Maine, and Vermont.[14]

In New Hampshire the first person known to have cultivated hops commercially was William Campbell, who moved from Wilmington, Massachusetts, to Bedford, in Hillsborough County, about 1800. Campbell brought his roots from Wilmington, successfully raised a crop, and expanded his yards. Neighboring farmers soon followed suit. As a result, Bedford became an early center, and Hillsborough County became the acknowledged leader in hop production in New Hampshire. In fact, the 1840 census returns confirm that Hillsborough produced more hops than any other county in the United States.[15] Bedford and the adjacent town of Milford became marketing centers for a valley about five miles wide, bordered on the south by the Souhegan River and the Nashua-to-Keene stage road, where

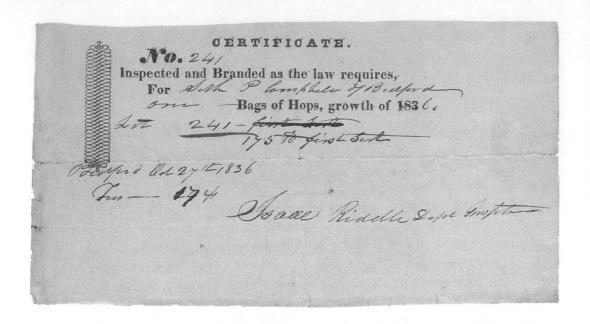

CERTIFICATE.

*No.* 241

Inspected and Branded as the law requires,

For *Seth P Campbell of Bedford*

*one* — Bags of Hops, growth of 1836.

*Lot* 241 — *first sort*

175 lb first sort

*Bedford Oct 27 1836*

*Ins — $174*

*Isaac Riddle Dep Inspr*

*1.2 Inspection certificate for Seth P. Campbell, one of a family of hop growers in Bedford, New Hampshire, dated October 27, 1836.*

*1.3 Oxford County, Maine, 1820. More than 80 percent of the hops produced in Maine at this time were grown in the Androscoggin River valley. (From H. S. Tanner, Map of the States of Maine, New Hampshire, Vermont, Massachusetts, Connecticut & Rhode Island, 1820. Courtesy of John W. Reps.)*

forty or fifty farmers specialized in hop raising. Hillsborough County continued to pilot the way in New Hampshire during the 1850s, followed by Merrimack, Stratford, Grafton, Worcester, and later Coos counties.[16]

Settlers in Maine followed the lead of farmers in northeastern Massachusetts and took up growing hops. In 1820, for example, nearly eighteen thousand pounds were grown in a very well-defined area of the state along the valley of the Androscoggin River. Over 80 percent of the hops in this state were harvested in five towns in Oxford County, namely, Brownfield, Buckfield, Denmark, Dixfield, and Hiram. Canton and Peru became the chief collection points, and Portland became the chief port for both foreign and domestic shipments.[17]

Vermont began the cultivation of hops comparatively late but made considerable gains in the 1840s. The craze swept several communities in the 1850s. Farmers were very successful with the crop in Windsor County, along the Connecticut and White rivers, and in Orleans County, where the valleys had the requisite well-drained soil.[18]

By the time the culture had spread northward to Vermont, however, the early centers in Massachusetts were no longer as active. The peak of hop production in Massachusetts occurred in 1836, when 847,590 pounds inspected in Massachusetts were sold. Thereafter the decline was sudden: 254,795 pounds were

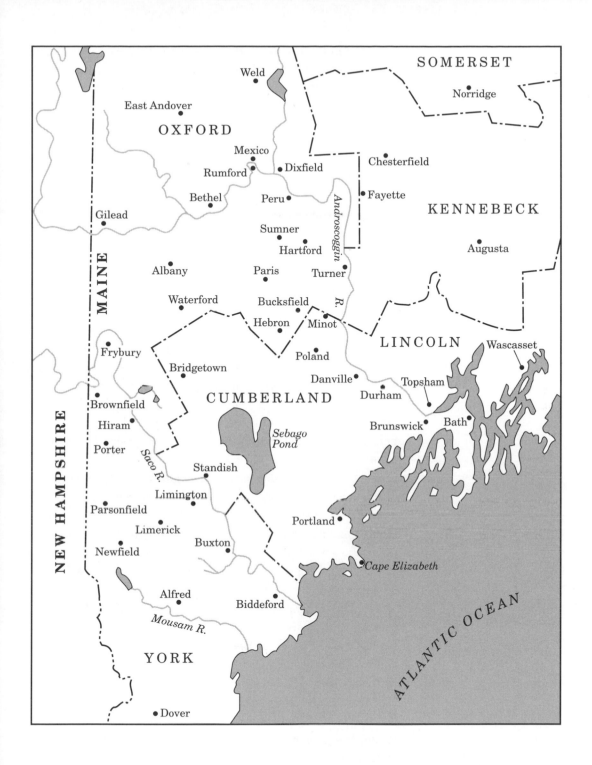

SOMERSET

Weld

Norridge

East Andover

OXFORD

Mexico

Chesterfield

Rumford

Dixfield

Fayette

Bethel

Peru

KENNEBECK

Gilead

Androscoggin R.

Sumner

MAINE

Hartford

Augusta

Albany

Paris

Turner

Waterford

Bucksfield

Hebron

Minot

LINCOLN

Frybury

Poland

Wascasset

Bridgetown

Danville

Topsham

CUMBERLAND

Durham

Brownfield

Brunswick

Bath

NEW HAMPSHIRE

Hiram

Sebago
Pond

Porter

Saco R.

Standish

Limington

Parsonfield

Portland

Limerick

Buxton

Newfield

Cape Elizabeth

Alfred

Biddeford

ATLANTIC OCEAN

Mousam R.

YORK

Dover

reported in the census of 1840, and only 121,595 pounds were noted a decade later.[19] In New Hampshire the peak came in the 1850s, when about 260,000 pounds were harvested; the census returns indicate that thereafter the crop diminished by at least half every ten years. By 1879 less than sixty acres was being grown, with a yield of less than 24,000 pounds. In Maine and Vermont the culture peaked after the Civil War and then rapidly declined.[20]

The reasons for the decline of hop culture in New England are not entirely clear, but economics and the general dissatisfaction among brewers and hop merchants with the inspection system played a part. By the late 1830s farmers in central New York were beginning to produce hops at a price that their brethren in New England found very competitive. Because hops drew so many nutrients from the soil, New England growers who did not heavily fertilize their yards experienced an inevitable decline in productivity, allowing New York farmers, who were cultivating new soil, to produce more for less. Further, although Massachusetts growers produced more hops in 1836 than in any other year, sales were conducted at an average price of seven and a half cents per pound, a great disappointment. The following year the average price dipped to six cents, and the market did not recover quickly.[21] Hence, there was little incentive to increase the quantity of hops grown.

About the same time, dissatisfied hop merchants began to voice their concern about the quality. In 1842, for example, eight Philadelphia buyers wrote to the Boston inspector, Benjamin Farnsworth, complaining about hops shipped from Boston and other Eastern ports. The shipments were labeled as first sort and were purchased by the consumer at the highest prices. Upon examination, however, the bales were often found to contain so many leaves and stems as to "render them unfit for the brewer's use."[22] Because they were having such difficulties, the Philadelphia brewers noted, the demand for hops raised in New York was increasing and they commanded a higher price. Another letter, signed by fifteen New York City buyers, warned ominously that unless immediate action was taken, the poor reputation of Massachusetts hops would put many growers out of business.

Meanwhile, the early successes experienced by the farmers of New England stimulated settlers who moved to upstate New York to take up the crop.[23] At the beginning of the nineteenth century, breweries in Philadelphia, Baltimore, New York City, and Brooklyn were beginning to supply the region with a considerable amount of beer. Could they depend upon the inland valleys of the northeastern states to provide them with hops?[24]

The need was obvious, and soon attempts were made to fulfill it. The commercial production of hops was introduced to central New York in 1808.[25] Credit for this development is given to James D. Cooledge, a native of the town of Stow in Middlesex County, Massachusetts. Cooledge purchased a farm about a half mile south of Bouckville, Madison County, and began his culture by securing a number of roots from his neighbors' gardens. Others nearby, such as Solomon Root, another native of Massachusetts, also had sizable hop yards. The principal difference between Cooledge and the other growers in the area

1.4 *Madison Township, Madison County, New York, 1853. The Cooledge and Root families lived near Bouckville, New York. After the pioneer grower James D. Cooledge died in 1844, his property passed to William, his son. (Anthony D. Byles*, Topographical Map of Madison County, New York, *1853.)*

seems to be that he was the first to transport a load of hops to the New York City market, in 1816. Thereafter, Cooledge was recognized as the earliest significant promoter, even though Root raised considerably more hops, and other early growers, such as Ezra Leland, became better known for their contributions to the agricultural press.[26]

By 1819 production in Madison County had reached such a level that, in response to the demands of growers, the New York legislature passed a law calling for the compulsory inspection and grading of hops, based on the Massachusetts statute enacted a decade earlier.[27] Two inspectors were appointed, one stationed in Albany and the other in New York City, and each sorted bales into first sort, second sort, third sort, or refuse. The annual reports of these inspectors suggest that, although Cooledge and Root may have been the first to recognize the lucrative rewards awaiting the commercial hop grower, other farmers in the area were not far behind. In Oneida County, for example, the first hop yard was begun in 1821 by Benjamin Wimble, an Englishman who had settled in the town of Sangerfield.[28] The culture seems to have spread eastward thereafter, to Otsego County. E. A. LeBreton, inspector of hops in Albany during 1828, reported that he had examined 1,263 bales weighing 277,502 pounds, raised in eleven counties, with Madison, Otsego, and Oneida foremost.[29] The average price was nine and a half cents per pound, somewhat below what could have been realized, because of the excess from the 1827 harvest and the depression being suffered by the brewing business.

As in Massachusetts, despite compulsory inspection, the quality of the hops that went to market was difficult to control. Of the 2,300 bales shipped through the Hudson River valley in late 1833, for example, perhaps only 200 bales should have received the designation "first sorts." Many hops were picked prematurely, and others were scorched or smoked in the curing process. By the early 1830s some Philadelphia brewers refused to buy upstate hops as well.[30]

Whatever the difficulties, New York's agricultural periodicals bolstered hop production by providing an exchange of information and reporting at length on the success of hop farmers.[31] By the time the 1840 census was completed, the combined production of all the New England states was only slightly greater than New York. More than 447,000 pounds were grown in the Empire State, almost double the produc-

TINGED WITH GOLD

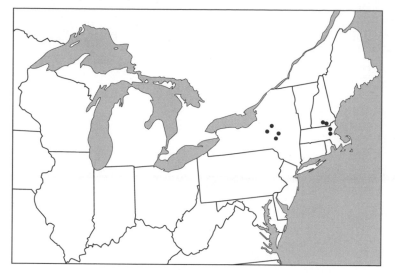

1.5 Commercial hop production in the United States, 1839. Each dot on the map represents 100,000 bales. The established fields of northeastern Massachusetts and southeastern New Hampshire are indicated, along with the newer plantings in central New York. (From Ruggles, Tabular Statements, 1874.)

tion of the next closest contender, Massachusetts.[32] Over the course of the next decade the production of hops in the United States nearly tripled, to almost 3.5 million pounds, and of this figure, over 2.5 million pounds was grown in New York.

About 1835 Otsego County took the lead from Madison as the center of production. These two counties were closely followed by Oneida, while Franklin, Ontario, Herkimer, and Schoharie counties would remain in a second group. Cooperstown was the principal hub at the eastern end, and Waterville was the chief node on the western end of this band. Hops were grown for miles in every direction around these villages, and the long freight trains, loaded with the baled crop, moved through the area with considerable frequency during and after the harvest. Were the growers proud of their crop? Gurdon Avery of Waterville stated that in 1842 he had raised 29,937 pounds of hops on twelve acres of land, and he bet $1000 that no grower in the world could equal the quality and quantity of his crop. The newspaper noted he was "doubtless safe; for his crop is unequaled by any one on record."[33]

After the Civil War, although prices fluctuated for a few years, New York's production continued to dominate the markets. The statistics reported in the tenth census of the United States (1880) reiterated the basic pattern that had been found decades earlier. Otsego County still held the lead, with 4,441,000 pounds; Oneida County had moved up to second place, with 4,075,000 pounds; and Madison County was third, with 3,823,000 pounds. Schoharie, Franklin, and a "late

1.6 *Bird's-eye view of Waterville, New York, 1885. Tepeelike collections of regularly spaced poles characterized the land-scape in central New York wherever hops were grown. This view shows the close relationships between the fields and the hop houses as well as the shapes of these buildings.*

bloomer," Montgomery County, were fourth, fifth, and sixth in production.[34]

At the peak of production in New York, in the early 1880s, the largest farms were those belonging to William P. Locke and John J. Bennett, both of Waterville, with 168 and 125 bearing acres, respectively. Such large fields were the exception, however, not the rule. Most hop yards were modest, four- to six-acre layouts, often situated in the relatively level bottom land that flanked one of the many flowing streams in the region.[35]

The tenth census of the United States indicated that New York was producing over 21 million pounds, or over 80 percent of the total crop in the country. By the time the results of the eleventh census were complete, however, New York was responsible for only 20 million pounds. The slight drop in production is not as significant as the fact that this represented only about 50 percent of the total amount grown. In 1900 another slight reduction is evident in the amount New York produced— 17 million pounds—but this constituted only 35 percent of the total production. The pattern of decline was clear, as competition from the new, larger hop fields being planted in the Far West was becoming ever stronger. The western fields were not only less expensive to operate but, more important, produced a larger crop per acre. The census of the fall 1879 harvest in the top six hop-producing counties of New York and California indicated that although New York led in the total number of

*1.7 View along Dayton-ville Road, near Water-ville, New York, c. 1900. (Courtesy of John Alden Haight.)*

TINGED WITH GOLD

pounds, California fields were over twice as productive per acre.[36]

Gurdon Avery's challenge did not go unnoticed in the Midwest. "Suckers, Hoosiers, Badgers, Wolverines, Hawkeyes, Pukes! Where are ye? What the East has done the West can outdo," or so it seemed to the *Prairie Farmer* in 1843.[37] This call from Chicago went almost completely unanswered, however, until after the Civil War.

Although a number of farmers in Illinois and Iowa experimented with the crop,[38] when midwesterners began to look for expertise in hop growing, invariably they turned to the growers in Wisconsin. Hops had been discovered growing wild by the first settlers in that area of the country. Those who had traveled near Racine in the mid-1830s remembered that they had found "many varieties of wild onion, plum and crab apple, and among them the hop."[39] The writer further noted that in 1852 or 1854 he had planted one of the wild hop roots in his garden and allowed it to grow "in regular old-fashioned New-England style." It first climbed up two fence posts and then crept along over the lattice of one of the hennery yards into the field.

Immigrants to the region brought a considerable amount of agricultural knowledge with them, so that hops were introduced in a number of areas. Generally speaking, the culture was introduced by settlers who emigrated from the Empire State.[40] For example, James Weaver, who had been a hop farmer in New York, settled near Sussex, in what became Waukesha County, Wisconsin, in the summer of 1837. Weaver apparently brought a number of roots with him and began a hop yard.[41] Indeed, the census returns for 1850 show that Waukesha County had a virtual monopoly on hop culture, producing 13,119 pounds of the total of 15,930 pounds harvested in the state.[42]

Sauk County, Wisconsin, also had its pioneers and eventually claimed to have become the center of hop production. In 1842–43 "the distinguished Hungarian political refugee," Count Agostin Haraszthy, "founded what is now Sauk City, where he planted the first hop yard in our State, and encouraged others to do likewise."[43] Apparently unaware of Weaver's efforts and the earlier work of Waukesha County growers, by the late 1860s most agriculturalists believed that commercial hop growing in Wisconsin began in Sauk County. There, Jesse

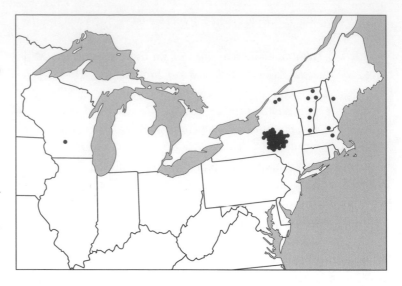

Cottington, an English immigrant who had managed the hop fields of C. C. Palmer in Waterville, New York, began a sustained effort in the culture by planting a few of the roots he had brought with him on his farm in the town of Winfield, in the spring of 1852.[44]

The spread of hop culture in Wisconsin seems to have been linked to a decline in wheat prices, the failure of the hop crop in New England, and the admirable soil and climate that the state offered. The rise of production in the Badger State can only be described as meteorlike. Registering only 133 pounds in the sixth census of the United States (1840), production over the next decade increased to 15,930 pounds. By 1860 it had increased over eight and a half times, to 135,587 pounds, and four years later it amounted to 385,583 pounds.[45] Farmers in the region who were asking for information in 1863 and 1865 caused the *Wisconsin Farmer* to look through articles on hop raising that had been published in the *Country Gentleman*, the *Genesee Farmer*, and Bridgeman's *Young Gardener's Assistant*, while sniffing that "we are frank to say we believe that there is nothing like pure cold water as a beverage, and that malt liquors of any kind, are, in our estimation, not simply miserable stuff, but productive of disease and unhappiness."[46] On the other hand, one of the most outspoken advocates of hop culture was the *Prairie Farmer*, which engaged Dell Pilot, a prominent grower in Kilbourn City, Wisconsin, to contribute a series of articles in 1867 and 1868. He was well placed, for according to the local newspaper, Kilbourn City, located on the east bank of the Wisconsin River, was "the greatest primary hop depot in the United States—perhaps the world."[47] The editors of the paper in this young town of just over fifteen hundred inhabitants might have been more modest in their claims had they been more familiar with markets in the East.

Up to 1868 Wisconsin farmers had not suffered from insect infestation, and they were spurred on by reports from England and Germany that the crops abroad had been destroyed by drought, mold, and blight. Redoubling their effort, they cashed in on the opportunities presented them.[48] As in the East, most hop farmers grew but a few acres. In the town of Fairfield, Sauk County, for example, ninety-five growers cultivated 438.5 acres, for an average of almost 4.5 acres each. Gardner Myers tended 30 acres, which was the largest field, while H. Kingston cultivated only a half acre, the smallest yard.[49]

The culture spread in Wisconsin from Waukesha, Sauk, and Dane counties to Jefferson and Columbia counties. In retrospect, 1867 was the peak year financially, for hops sold at fifty or sixty cents per pound, and 1868 was the best production year in Wisconsin. At that point, over six thousand acres were under cultivation in Sauk County, the leader in the state.[50] By the end of the season, however, because of an especially good harvest elsewhere, the price had dropped considerably. The figure of seventeen cents per pound was widely quoted, an amount that frightened most farmers and led them to consider plowing up their fields. Hop growing was likened to mining for gold in California and oil speculation in Pennsylvania. "Anything that amounts to speculation and is extensive enough to call in large numbers of people, is always a curse to the people who engage in it, and generally to the region where it is carried on."[51] By mid-November 1868, the *Wisconsin Mirror* frankly urged hop farmers to "Plow Them Up!"[52] Many first-time growers did. Was the hop business being overdone? Many growers did not want to face the question.

Although by 1870 the production level had climbed to 4,630,155 pounds, the peak had been passed. By the early 1870s the hop craze in Wisconsin was all over.[53] Only 1,966,827 pounds were harvested in 1880. In another decade it became clear that Wisconsin's fall was as dramatic as its rise. New York growers, relieved at this loss of competition, nevertheless eyed the far western states with growing consternation. As A. R. Eastman, a well-known grower from Waterville, noted in 1890:

"None of us are so blind that we cannot see the menace to our industry which is rapidly looming up on the Pacific Coast." The rich, alluvial bottom lands of California, Oregon, and Washington seemed especially well adapted to hop growing, and the plants were seemingly immune to insect infestation and plant disease. "Now with these facts plainly in front of us," Eastman continued, "and the history of hop growing in Wisconsin behind us, what shall be our course for the future?"[54]

Whatever solace New York growers might have gained in the failure of the hop culture in Wisconsin, it was short-lived. The states of the Far West continued to expand their production capabilities, while New York grew fewer hops with almost each passing year. From a crop of slightly over 17 million pounds recorded in 1900, production dropped to 8 million pounds ten years later, and after World War I only 724,000 pounds were grown. The future for hop culture lay far to the west.

Almost everywhere that easterners settled on the West Coast, the cultivation of hops followed. In California, although mining interests were the first to claim attention, as more fortune seekers poured into the state, the need for provisions became crucial. Many would-be miners or merchants returned to the occupation they knew best, farming. Further, the more enterprising farmers began to experiment with a wide variety of crops and soon demonstrated the productiveness of the soil with vegetables, fruits, and grains.

Wilson G. Flint has been credited with introducing hops into California.[55] In 1849 Flint moved to San Francisco from New York City, where he had been involved in the mercantile business. Wilson Flint was apparently soon followed by his brother, Daniel, who became a Sacramento agriculturalist and earned a reputation as dean of the California hop growers. The Flints were natives of New Hampshire, where their father had been a farmer in the Ashuelot River valley. Remarkably enough, however, the family had never raised more than one or two hop plants in their kitchen garden.[56] In 1854, with the returns from his importing business beginning to wane, Wilson Flint turned to farming. Daniel Flint later recorded his brother's experiments, stating that the first imported hop roots came from Vermont in the winter of 1855–56.[57] Daniel himself began commercial production of hops in the spring of 1857, when he purchased sixteen acres of the bottom land along the Sacra-

TINGED WITH GOLD

mento River, about one and a half miles below Sacramento. Having obtained roots from Wilson, he planted a trial yard, built the first hop kiln, and constructed the first hop press on the West Coast. In later years, to underscore his role as precursor, he stamped all the bales from his yard with his trademark, "The Pioneer Hop Grower."[58]

Another early pioneer in the importation of hop roots was Amasa Bushnell. Formerly a hop grower near Syracuse, New York, Bushnell brought roots with him. In 1856 he planted them on a ranch in San Mateo County, on the site of the present city of Burlingame, and two years later he teamed up with a Syracuse native, Samuel Dows, to set out more hops in Vine Hill, on the east side of Green Valley in Sonoma County. The Vine Hill field yielded a harvest of two hundred pounds.[59] In 1860 the visiting committee of the state agricultural society reported that the fifteen-acre hop field of Bushnell and his new partner, Otis Allen, was the largest in the state, and subsequently they were awarded a silver vase at the state fair for the best acre of hops.[60]

In the spring of 1863 the California legislature encouraged all agriculture and manufacturing in the state by passing the so-called State Bounty Law. This promised $1,000 to the grower who produced the first 1,000 bales of 200 pounds each, and for the same quantity produced in the first, second, and third succeeding years, $600, $400, and $200, respectively. Two years later, Daniel Flint collected a reduced premium of $250 for his crop of over 250 bales.[61]

Throughout the 1860s California brewers continued to rely on eastern hops, often mixing in locally produced hops in their wort. Dealers and importers, not eager to cut into their own profits, perpetuated the opinion that New York hops were superior and forced California growers to sell for less.[62] In the early 1870s, however, brewers of San Francisco and Sacramento gradually came to realize that the California hop easily rivaled the eastern product in freshness and quality, and local hop growers began to prosper. The demand for roots increased, so that prominent growers, such as Daniel Flint, soon reported the sale of thousands of dollars of cuttings from their yards.[63]

By 1867 the total hop crop in the state of California was approximately 425,300 pounds.[64] In the same year, California hops suddenly captured a worldwide reputation when, at the Paris Exposition, the first premium for porter and ale was

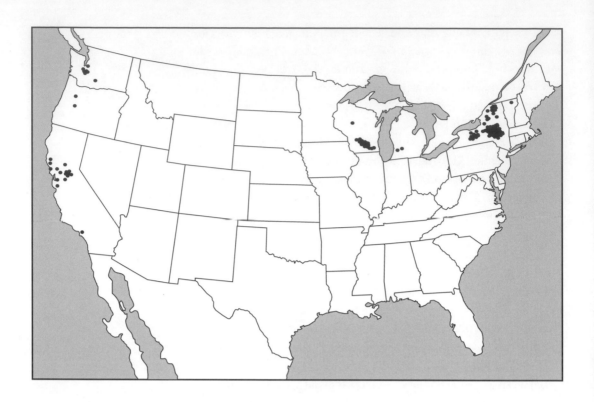

awarded to the Smith Brothers of New York. When the judges asked the person in charge of the Smith Brothers display how and of what they managed to make their product, the reply was "Croton Water, with California hops and barley."[65] At the 1876 Centennial Exposition in Philadelphia the premium for the best hops was taken by a crop from St. Helena, in Napa County.[66]

Although experimentation with hops was widespread, the broad, fertile Sacramento Valley, lying between the Coastal and Sierra Nevada ranges, became the early center of production. In the years immediately after the Civil War, Sacramento County produced about half the entire harvest for the state. Because Sacramento was the center of the state's agricultural press, it disseminated much of the available technical information and continued to foster experimentation and improvement. Some of the best-known growers were Glacken and Wagner, Ezra Casselman, W. H. Leeman, Lovedale Brothers, Palm and Winters of Walnut Grove, Flint and Raymond, W. F. Warburton of Elk Grove, and George Menke.[67]

Mendocino and Santa Clara counties were generally second and third in the number of acres and in the amount produced.

Ukiah and Hop City became the principal northern trading centers and continued to figure prominently in markets outside the state. Although a number of new plantings in the early 1870s near San Jose catapulted Santa Clara County into first place for a few years, the Bay Area was never a serious leader in production.

Once again, the early hop fields of California were not especially large. In 1874, for example, the largest field in Mendocino County was a little over twenty acres. The typical growers in the Ukiah and Sanel valleys would set out between seven and ten acres, producing between eight hundred and two thousand pounds per acre,[68] and a single unsuccessful crop might easily lead a farmer to conclude that his attention was better spent cultivating grapes, apples, or peaches. The drop in prices which occurred in the mid-1870s led to a call in New York and Wisconsin to cure over-production by cultivating only half the potential yield. In California growers were producing other crops in addition to hops, and they were advised not to plow up their fields but not to expand them either. West Coast farmers could afford to take a chance and wait for the price to rise.[69]

The outlook did not change appreciably for the next four or five years, but when the economic situation began to show promise, some of the large fields were planted. These fields dominated the industry over the course of the succeeding decades.[70] While the hilly, narrow valleys of the Mendocino-Sonoma hop district restricted the size of the fields, the inland plains and the flat river bottoms were nearly ideal. In 1883 the physician D. P. Durst planted his first hops in the broad fields bordering the Bear River, west of Wheatland, in Yuba County. Durst increased his acreage until he became known as "the world's largest hop grower," and by 1886 he had 100 acres under hop trellis, producing 2,000 pounds to the acre. By 1892, he had increased his holdings to 240 acres and owned all the land along the Bear River for five or six miles.[71] The size of Durst's operations was impressive, easily rivaling the largest farms of the East. Even Durst's efforts were quickly surpassed, however, for by 1895 the Pleasanton Hop Company, in the Livermore Valley of Alameda County, had 400 acres in bearing.[72]

The Durst plantation was one of several in southern Yuba County that established themselves in the forefront of production and remained there for nearly a half century. In fact, in

1.10 Willamette River
valley, 1878. The early
centers of hop production
were located between
Salem and Albany. (From
J. K. Gill & Co., Oregon,
1878.)

1889 and 1890 Yuba County, with an average of 2,340 pounds per acre, had the highest yield of any county in the nation.[73]

From 1915 until 1922 California was the leading hop-producing state in the Union, responsible for over 50 percent of the total U.S. production in some years; the state's acreage peaked in 1916.[74] Whereas in the East or in England, growers might complain that the crop never fully matured from lack of sunshine, too much rain, or mildew, in California this never seemed to be the case. Further, as late as 1891 California was free of attacks from insects that were common in the Pacific Northwest and the eastern United States.[75]

As elsewhere, however, the decline was precipitous. California growers were some of the hardest hit by Prohibition; only ninety-three farms were producing in 1929.[76] Nearly a decade later, only three counties—Sonoma, Mendocino, and Sacramento—were even modestly important, leaving the leadership in the industry to the states in the Northwest.

Hops are known to have been grown for domestic use in a number of locations in Oregon Country in the first half of the nineteenth century, but not until the second half are the crop records clear.[77] Nearly five hundred pounds were recorded in the 1859 census—that is, approximately two and a half bales were produced—so that at least one or two farmers must have recognized their importance. The first grower to claim the distinction of having planted the earliest commercial yard in Oregon was William Wells of Buena Vista. Wells began production in 1867, but his fields remained small and his activity did not inspire much emulation.[78]

Other experiments were tried about the same time. In 1867 Adam Weisner emigrated from Wisconsin to Oregon and settled at Buena Vista. He rented five acres of the upland and planted it in hops, having brought the roots with him from Wisconsin. Although he had gone to considerable expense and built a hop house, Weisner's attempt proved to be a failure. He sold some roots to George Leasure, who planted them in the spring of 1869 in the Willamette River valley near Eugene. This was the beginning of a yard that bore hops for over thirty years and was often referred to as the earliest in the state.[79]

Not until the mid-1870s did the activity of a few hop growers begin to be noticed. In November 1871, for example, the first prize premium for hops at the Oregon State Fair was won by Mrs. T. W. Thompson, for one pound. The prize was listed along

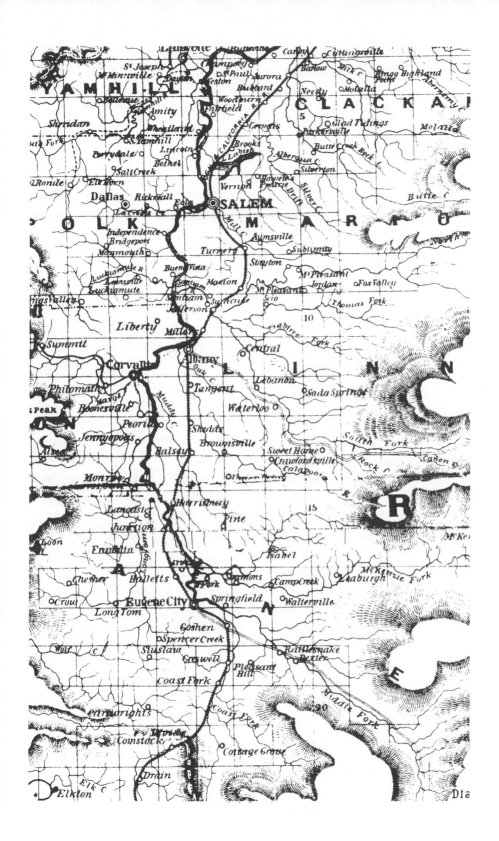

with others, under seeds.[80] Soon managers of the state fair were strongly urged to encourage the state agricultural society in fostering hop growing, as had occurred in eastern states. Progress was still slow, however, for as late as 1876 William Wells claimed to be personally acquainted with most of the growers in the state.[81]

The Willamette River valley, stretching from the south to the north behind the Coastal Range, became one of the most active hop-growing regions in the United States. The crop was raised on the river bottom and on the shelves on either side above the flood plain. The early fields were modest in size. Farmers found that if they planted hops along with their other crops, they needed only to employ local labor and could take advantage of favorable market conditions to export their crop. San Francisco was the closest market. Messrs. Hayden and Lincoln, hop dealers of that city, offered a first prize of twenty dollars in gold at the 1876 Oregon State Fair for the best ten pounds of hops grown, cured, and pressed in Oregon.[82]

Oregon's production statistics were very promising during the late nineteenth century. In the tenth census (1880), Lane County was clearly the leader, with 143 acres under cultivation, producing 99,298 pounds. A local history published four years later remarked that "the crop runs from fifteen hundred to three thousand pounds per acre."[83] Marion, Linn, Clackamas, and Polk counties were in the next class, although each had an average of only about 30 acres in bearing.[84]

The story of the rise in Oregon's commercial production and the shift in the geographical center of the fields can be traced in the prominent Seavey family. Alexander Seavey, a sailor, pack-train driver, and pioneer merchant of Lane County, took up farming in 1855. Seavey began raising hops about 1877, and by the turn of the century had one hundred acres in production near Springfield. More impressive, however, were the holdings of his three sons, John, Jesse, and James. The last, in particular, became well known as a Portland-based dealer and owned five ranches: in Forest Grove, Washington County; Oregon City, Clackamas County; Eugene, Lane County; and two in Corvallis, Benton County. The newer fields were laid out downstream, to the north.[85]

By the turn of the century the four principal marketing centers were Eugene, Salem, and Independence, all on the Willamette River, and Grant's Pass, on the Rogue River. In-

dependence claimed the title of "Hop Center of the World," largely due to the enormous fields at "Bird Island," made up of the nearby Horst and Hirschberg ranches, and "newsy places in their vicinity."[86] Lane County was left behind, because Marion and Polk counties soon contained over half of Oregon's acreage and their share increased to three-quarters by 1919. In Polk County about twenty-five hop houses were erected in the 1894 season alone.[87]

These counties thrust the state to the forefront. Oregon ranked first in the production of hops in the United States from 1905 until 1915 and again from 1922 until 1943, when it was superseded by Washington State. The number of Oregon growers doubled from 1919 to 1929 and by developing foreign markets were able to plant their peak acreage in 1935, when nearly 26 million pounds of hops were harvested. By 1939 approximately half of all American hops were being grown in Oregon, and the other half was about equally divided between Washington and California.[88]

About this time a remarkable change took place in the nature of the industry. A dramatic reduction in the number of hop

*1.11   Commercial hop production in the United States, 1899. Each dot on the map represents 100,000 bales. By the turn of the century, the growth of the Pacific Coast states eclipsed the long-held leadership of New York. (Source: William R. Merriam, Director,* Census Reports, Twelfth Census of the United States, Taken in the Year 1900 *[Washington: United States Census Office, 1902], vol. 9, pp. 517–19.)*

growers occurred, along with a drop in the number of acres devoted to the crop. The latter decrease was coupled with the conversion from manual to mechanical picking, which put the small growers at a disadvantage. Oregon growers soon had to acknowledge they were losing their edge to the larger commercial growers in Washington.

The credit for being the first to promote commercial hop growing in Washington Territory generally goes to Jacob R. Meeker. A brewer in Olympia, Charles Wood, furnished Meeker with roots and encouraged him by promising to buy the hops. The roots were planted in the spring of 1865 near Sumner, in what became the great hop-growing area of the Puyallup Valley. The first crop yielded a single bale of 185 pounds which, when sold in Olympia for 85 cents per pound, brought $159.25. This was more than enough to encourage Meeker and other pioneers to increase their efforts.[89]

There is convincing evidence that this story, well known to those who subsequently read Meeker's self-serving publications, can be challenged by the deeds and words of other early growers. In the spring of 1865, L. F. Thompson and E. C. Mead set out twenty-five hundred roots they had shipped from Sacramento, California, most likely having purchased the plants from one of the Flints. This could account for Daniel Flint's boast, "Most all of the hops on the Pacific Coast came from my yard."[90]

Regardless of who was the first to cultivate the crop, the rich alluvial soil of the White River valley seemed to have been custom made for hops, as growers harvested crops supposedly unheard of anywhere else in the country. By 1874 eighteen growers in Pierce County reportedly had a total of about 150 acres in hops. The largest firm was Thompson and Mead, with about a third of the total area under cultivation. Most of the early growers—J. P. Stewart, J. F. Kincaid, J. R. Dickinson, G. H. Ryan, and H. C. Helmhold—began with only one to four acres, expanding in the late 1870s.[91] Puyallup was the earliest marketing center, largely through the efforts of Ezra Meeker (Jacob Meeker's son), and J. P. Stewart, who convinced the owners of the Northern Pacific Railroad to include the town on the line from Tacoma to the coal mines at Carbonado. Puyallup teemed with about three thousand native Americans and an equal number of white pickers by the early 1880s.[92]

From the fields of Pierce County, the craze spread in every direction. To the north, in King County, in 1868, the Wold

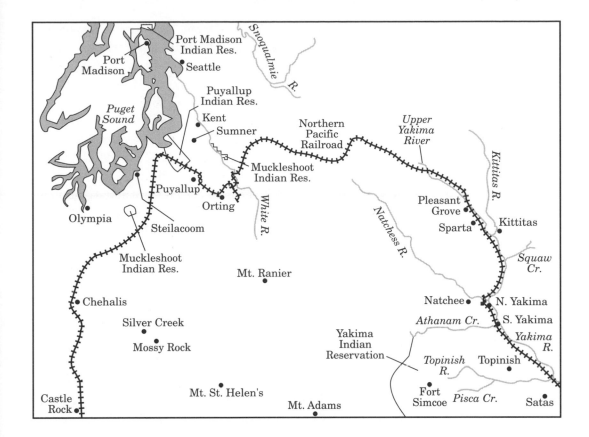

brothers purchased two thousand plants from Ezra Meeker of Puyallup, and carried them by mule to their farm in the Squak Valley. Their success was almost immediate, and the area devoted to hops increased rapidly.[93] In 1883, for example, the area in Pierce County was estimated to be 873 acres, an expansion of 40 percent over the previous year. In King County, however, the increase was even larger, and the fields under cultivation were estimated at 800 acres. In part, this was due to a single farm of 300 acres, owned by the Seattle Hop Growers Association. In the late 1880s and early 1890s, King County was also among the acknowledged leaders in the state, producing about twenty-one hundred pounds per acre.[94]

The settlers on the eastern side of the Cascade Mountains were quick to notice the phenomenal success of farmers in the Puget Sound area. Agriculturalists in the Yakima River country, between the Cascade Mountains on the west and the Columbia River on the east, were comparatively isolated but found that hops were one of the few crops that they could raise profitably.[95] Whereas the price of hops was high enough to

*1.12 Principal hop-growing areas in Washington Territory, 1881. The Puyallup and White River valleys south and east of Tacoma, and the Yakima River valley further to the southeast proved most productive. (Source: Lieut. Thomas W. Symons,* Map of the Department of the Columbia, *1881.)*

cover all the expenses of raising them, the cost of transporting any other produce was often greater than its market value. Charles Carpenter is generally given credit for having begun the culture in 1872. His first roots came from his father's farm in Constable, Clinton County, New York.[96] Yakima Valley growers shipped eighty bales of hops westward in 1876, and while they enjoyed a moderate increase in acreage in the following decade, after the Cascades branch of the Northern Pacific Railroad reached the area their future success was almost assured. In 1886 the principal farms were in the Ahtanum Creek valley, some twenty-five miles long and five miles wide, west of Yakima.[97] By the early 1890s the culture spread to the southeast of that railroad town, and Yakima County was recognized as the principal hop-growing area of the state. In an area known for its hot, dry summers, considerable attention was devoted to artificial irrigation, but the crop flourished in the heat. Toppenish and Moxee City became major market centers; the latter has remained so to the present day.

As in other parts of the country, some of the earliest growers became some of the best known. One of the greatest hop enterprises in Washington during the late nineteenth century was the Puyallup Hop Company, in the White River valley, near the town of Kent.[98] Controlled by the Meeker family, the company's operations included growing and dealing in hops on an international scale. In 1883 Ezra Meeker took his first lot of five hundred bales to London, while in 1884 he sent fifteen hundred bales and in 1885 about eleven thousand bales. During the 1890 season his company shipped six solid trainloads of hops to London. Together with their other shipments, this represented one-fifth of the entire harvest for the state. In fact, Ezra Meeker, by virtue of his writing and various public positions in the Puyallup Valley, came to be regarded as the ranking expert in the Northwest. He served as judge of hops at the Columbian Exposition in Chicago, in 1893.[99]

At the same time, the largest hop ranch in Washington, and perhaps the largest hop ranch in the world, was owned by the Snoqualmie Company, situated in a natural prairie, surrounded by a dense forest in the Snoqualmie River valley, west of South Bend, in Pacific County. The company began in 1885 and planted 200 acres in hops, profiting handsomely, so that by 1890 it had expanded to 310 acres, harvesting about three thousand bales of two hundred pounds each.[100]

TINGED WITH GOLD

As might be expected, Washington growers also encountered problems. The extremely low price of hops in 1877 discouraged many of the hop growers in the Tacoma area. Some solace could be gained when they learned that, by comparison with the New York and California crops, the hops produced in the territory were superior.[101] Although the region was at first untroubled by the insects and blight that had so complicated the lives of Eastern farmers, by the late 1880s and especially in the early 1890s, an alarm spread throughout the Northwest and California as the hop aphis appeared without any warning. Various remedies were tried, but to no avail. Whole fields were infested.[102]

Because of the increasing number of disputes between growers and buyers who paid in advance for given amounts, in 1899 growers in Washington were the first in that area of the country to become subject to state legislation that appointed a state hop inspector. Henceforth, if there was a disagreement, the inspec-

*1.13   Jacob Meeker and his crew with mule team, Puyallup, Washington, 1888. Meeker, standing to the left, was the pioneer grower in a family that played an important part in establishing and promoting hop culture in the Pacific Northwest. (Washington State Historical Society, Tacoma, Wash.)*

tor determined the grade and quality of the crop. Joseph N. Fernandez of Puyallup was appointed by the governor as the first inspector.[103]

Early-twentieth-century production in the state was not exceptionally large by western standards. The increase was steady, however, reportedly growing from 1,129 acres in the 1920 census to 2,814 acres in 1930, and a little over 4,600 acres in 1940.[104]

Perhaps the most important lesson twentieth-century hop growers have learned is their near-complete dependence upon the international market. The flow of American hops increased when the production in Europe was slowed. A small European crop in 1911 greatly enhanced the demand for the American product, and the price went up. World War I altered the price structure even further, so that by 1919 U.S. hops were selling at a high of a dollar per pound. The hostilities on the other side of the Atlantic meant that European hop fields were not being cultivated. If growers were interested in making money, the international market was the place to do it.

Although the export market was strong, domestic sales lagged. Adverse legislation was having an effect on brewing in this country by 1913, as the Anti-Saloon League and its advocates sought to ban not only distilled liquors but also wine and beer production. By 1916 twenty-three state legislatures had adopted local prohibition laws. As an increasing number of counties and municipalities went "dry," hop growers found that their prosperity was clouded in an air of increasing public hostility. Prohibitionists were given an opportunity to press their cause when the United States entered World War I. With the U.S. troops abroad, in 1917 the Food Control Bill was passed, forbidding the use of foodstuffs for the production of alcoholic beverages. Hop growers, anticipating the drop in demand and increase in shipping difficulties, took acreage out of production. The adoption of the Prohibition Amendment in 1919 all but completely halted the domestic production of beer and consequently forced hop growers in the United States to rely completely upon markets outside this country.[105]

Immediately after World War I, the export demand for hops far exceeded the supply, and American hop growers received top prices for their crops. New facilities were built and new fields set out. German and Austrian hop growers, unable to rebuild their fields rapidly enough, had no choice but to allow

Pacific Coast growers to dominate the markets.[106] As European growers regained their preeminence, however, the question of what useful purpose the overexpanded fields, equipment, and buildings might serve became more evident. Several large fields were plowed up and set in fruit trees. In California some kilns were turned into berry, fruit, or fish dryers.[107]

The resolution for the Twenty-first Amendment of the United States Constitution, reversing the Eighteenth, was ratified in 1933. As beer production increased almost overnight, the demand for hops surged and the price jumped accordingly. Many farmers responded by increasing their acreage. Efforts made in New York and Wisconsin to reintroduce the crop repeatedly failed; only the hop-growing regions on the West Coast remained viable. Oregon growers took the lead, expanding their fields to nearly 30,000 acres. California fields, at a low of 3,300 acres in 1930, rose to 7,800 acres in 1934, while the fields in Washington were increased to a lesser degree. This increase in acreage led to a surplus, and the price plummeted in 1935.[108] Faced with a few years of continued economic difficulty, growers appealed to the federal government for help in regulating production. The Hop Marketing Agreement and Order established the Hop Control Board in 1938. It did not market hops but rather attempted to control surplus production and prevent unwanted sales. Trouble in Europe once again interrupted the international market, production worries diminished, and the board was terminated in 1945.[109]

Other factors also influenced the domestic market. After the repeal of Prohibition, while the amount of beer consumed rose and beer making also increased, tastes had changed. The "good old lager beer," aged for months with the aid of a high pound or more of hops to the barrel, was becoming a drink of the past. The public seemed to prefer a less bitter beer, and thus the hopping ratio, the amount of hops used to flavor a barrel of beer, steadily declined.[110]

Because growers on the Pacific Coast had expanded their fields during World War II, when peace came and European producers recovered, the market glutted once again in 1948 and 1949, and prices fell. The majority of the growers favored another federal marketing order, which was reinstated, but both production and prices were affected only modestly by this and subsequent attempts at regulation.

Oregon slipped from first to second place by 1943, and

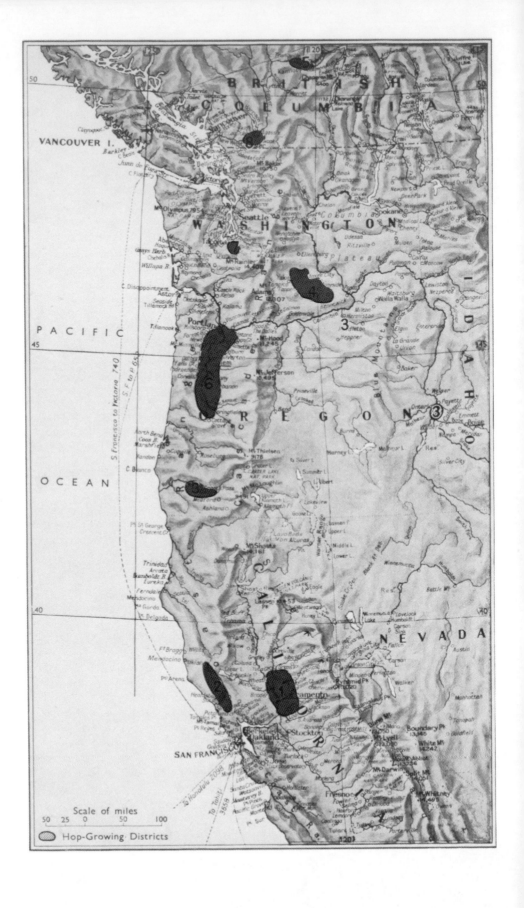

Hop-Growing Districts

dropped to third place by 1953.[111] Attacks of Downey mildew were blamed for the withdrawal of land from production. Washington, on the other hand, rose to prominence, so that by 1963 it accounted for more than half of the total U.S. production; seven years later, it accounted for 70 percent; and in the 1980s, about three-quarters of the nation's crop was produced there, all from the Yakima Valley. As in Oregon, California fields have dwindled from their high point in 1951 of 9,500 acres to nearly nothing, as the remaining fields in Mendocino and Sonoma counties have been reduced to extinction. In 1960 new fields in Idaho edged out Oregon for third place.[112]

This review of the first two hundred years of commercial hop growing in the United States has defined the principal regions in which the crop has been grown. Each region has seen similar developments and shared in a common agricultural tradition. Repeatedly, the early pioneers took advantage of the new soil and attempted to maximize their profits. The agricultural press, searching for information about the crop, turned to their better-informed brethren and carried the news of the financial benefits that were being reaped elsewhere. Both successes and failures were seen. In each region the decline was largely due to the economic conditions that influenced both the domestic and international markets for hops, although plant diseases also encouraged the abandonment of the vines. In each region, however, the growing and harvesting processes were similar, and the concerns of the grower and the picker were more immediate.

*1.14  Principal hop-growing districts of the United States, c. 1950. (From Anglo-American Council on Productivity, The Hop Industry, map 2.)*

# ❧ 2 ❧
# Growing and Harvesting

To understand the possibilities for profit as well as the difficulties associated with raising hops, it is necessary to know the steps involved in growing and harvesting the crop. After the farmer decided to raise hops, he had to choose the land, prepare the soil, lay out the field, set the roots, cultivate, dress and pole the vine, and when the time came, pick, dry, press, and bag the crop for shipment. These steps were slightly different from one area of the country to another, and were improved upon with time and experience. Among the numerous experiments tried by hop growers, some were aimed at training the vine in such a way as to improve the quality of the crop while making the job of picking quicker and easier. Others attempted to improve the drying or pressing equipment. The majority of the attempts at innovation were devoted to decreasing the expense of handling the crop at harvest time. However, much needed to be done before then.

The first requirement for hop growing was to obtain a suitable area on which to cultivate the crop. A reasonably level piece of ground with a deep loam provided the best situation. The levelness of the land ensured that the mellow soil would not wash away or gully, disturbing the roots of the plant. On the other hand, it was also important to avoid a sunken area where the ground was too wet. A slight slope or a low hill with a broad valley, well watered, was preferable because it offered the tall vines protection against the winds of summer storms. The southern side of the gentle slope was favored because the route of the sun and the direction of the wind would assist the farmer's efforts. Most of the hop-growing regions of the East and West coasts enjoyed these natural advantages.[1] In some cases growers selected locations that could be modified by the introduction of tree belts.

As with many other field crops, hops require a rich, fertile soil. Early pioneers worried little about this, but they soon learned that the vines took a considerable toll on the soil. New land, rich in organic matter, was quickly exhausted, and fertilizing became an important concern. Barnyard manures, particularly cow and horse manure, were spread on the surface, plowed deep, and subsoiled, to provide the roots with more nutrients and encourage them to penetrate to a greater depth, as much as three or four feet. In addition to animal manures, ashes and plaster were sometimes incorporated, as was muck, composed of the leaves of trees and other vegetable matter.[2] In

2.1 *The James Seavey Ranch, north of Springfield, Oregon, c. 1914. The ranch house and barns can be seen on the left; the kilns and cooling houses are evident on the right. (Lane County Historical Museum, Eugene, Oreg.)*

established hop fields, manuring followed the harvest season, not only to ensure well-fertilized roots, but also to protect them from freezing during the winter.

Planting a hop yard required the farmer to lay out the ground in rows, in the form of a grid. With a horse and light plow, the farmer made furrows six to nine feet apart north-south and perhaps eight feet east-west.[3] Next, a marker or stake was placed at each intersection of these lines to indicate where the roots were to be planted, in small mounds or hills of earth. Another variant was to plant the roots at five-foot intervals, with one pole to each hill. In the first case, approximately 770 hills per acre were formed, with two poles to each hill and two vines to each pole. In the second instance, approximately 1,700 hills were formed, each with two vines. This meant not only that more sets had to be planted and more poles had to be purchased, but also that more manure had to be spread. Thus the second method, involving more of the farmer's resources and attention, required a much greater financial investment.[4]

Although hops were known to have been begun from seedlings,[5] the vast majority of yards were laid out using root stock, a much quicker method. The initial sets were garnered from the roots of the previous year's growth on an established root. In the spring established roots sent out runners, which appeared just below the surface of the soil. The farmer grubbed out the runners, sometimes by digging with a hook similar to a potato hook. By cutting the runners into pieces about six inches or a foot in length, he made sets with two pairs of eyes in each section. These were usually planted in the early spring (as early as February in California, in April and May elsewhere), although they could be planted in the fall, generally in October. The sets were placed into the ground at the chosen intervals, in various fashions. Early-nineteenth-century directions would specify as many as five sets planted in each hill, but after the Civil War there seems to have been general agreement that the number could be reduced to one, two, or three, to allow only the strongest vines to mature.[6] One of the most commonplace late-nineteenth-century arrangements was a triangular formation of sets, about an inch or two below the surface. The roots were then buried in a slight hill, the eyes on the set pointing upward, to develop into shoots.

In the East no hop crop was expected in the first season, as the vine required a year to get firmly rooted. Often the hop field

TINGED WITH GOLD

was planted with another crop, such as corn or beans, care being taken not to disturb the hills of the hop roots. Both crops were cultivated simultaneously. In the West, because the virgin soil was so rich and the climate so favorable, a respectable crop could be grown by the end of the first year. At harvest time, any mature hops were collected, all the vines were cut and removed, and any hills where shoots did not mature were marked so that new plants could be supplied.

Hops are dioecious; that is, the male and female plants are different. As long as hops are raised from root cuttings, pollenization of the plant is not required for propagation. The female plants produce the flowers that become cones, which are harvested. At the base of the petals is the lupulin that is so desired as a brewing additive. A male plant, if allowed to grow in a hop field, will pollenize the females, which will then bear a larger "seeded" cone. Larger cones result in greater yields, but because the whole cone is harvested, dried, packed, sold, and transported, brewers did not want to pay for the heavier seeded cones. They preferred "seedless" hops, those with few seeds. Hence, the number of male plants was generally re-

*2.2 Male (left) and female (right) hop plants, as they develop near the top of the main bine. (Courtesy of John Alden Haight.)*

stricted to one in every one hundred vines. During the nineteenth century, the male vines were allowed to run up a pole five feet higher or more, so that the pollen could be blown in with the wind.[7]

The grower could choose any of several varieties of hop, either of European origin (*Humulus lupulus*) or native to America (*Humulus americanus*). The English Cluster and Grape were generally found to do well in New England and New York, whereas the Pompey and Red Bine were thought less productive.[8] Wisconsin followed the Eastern preferences. On the West Coast, English Cluster predominated in most fields, although other varieties were also grown and harvested at slightly different times to extend the picking season.[9] Growers seeking a more mildew-resistant strain introduced the Fuggle from Great Britain in the early twentieth century, and experimentation with these and other varieties continues.

The roots were dressed in the spring, as soon as the ground was dry and before the plants began to sprout branches. In New England this occurred between the middle of April and early May. The first step was to remove the manure or compost

*2.3   In the Fuggle* (left), *the cones are concentrated in the bunch; in the Early Cluster* (right), *the hops are larger and spaced in the bunches. (From Anglo-American Council on Productivity,* The Hop Industry, *plates 26 and 27.)*

TINGED WITH GOLD

used to protect the root during the winter and to cultivate the hill. This encouraged the crown of the root, which would develop into the vine, and discouraged the growth of runners, which would issue from the base or lower part of the crown and travel in a nearly horizontal direction, reaching the surface at various distances. These small shoots would, in fact, climb almost anything and bear fruit. If left unattended, they would usurp the place of the main root, eventually sapping it of all life.[10] Trimming involved opening the crown of the hill with a pronged hoe, removing sets or runners for future use, and carefully covering any young, exposed buds to shield them from a late frost.

More cultivation took place after the runners had been removed and fresh manure applied. In a small yard this was done by hand. In large fields a plow was run through the middle of the row in such a way as to turn the soil alternately toward and away from the plants. The idea was to fertilize the roots without disturbing them. In the larger fields manuring was a major undertaking, requiring tons of fertilizer.[11]

After the hop roots had been trimmed and cultivated, the next major task was to set the poles, or, in later years, set the poles and string the wires. Poling took place in late April and early May. Poles were twelve to sixteen feet long; some farmers recommended that they be as long as twenty-five feet.[12]

Growers selected a variety of woods for their poles, depending upon availability, quality, and price. Hardwood poles lasted three to perhaps five years. Spruce poles were mentioned as preferred in Vermont, although hemlock poles were also used and cost about the same amount. Cedar poles were a favorite throughout the country for their durability. They might allow ten or even twenty years' service, but they were also the most expensive, costing about twice as much as the other varieties. Wisconsin growers used oak or poplar poles. In California willow poles were most common, although they would last only three or four years. Spruce, madrona, and redwood were also employed.[13]

Wherever possible, poles were reused. After the harvest in the East, workers stacked the poles at the edge of the field in large tepeelike cones so that they could season properly and shed the rain. In the West the poles were often taken off the field and stored, furthering their life.

Because reused poles needed to be sharpened every two or three years, they soon shortened and were replaced. Damaged poles were discarded, for it was unwise to risk problems. In fact, one of the greatest fears at harvest time was that the weight of the crop, if soaked by a summer storm, would break a number of poles. These would rip through the vines and bring down adjacent plants.

The demand for poles created an almost continuous need for young trees throughout the hop-growing regions of the country. As hop culture increased in any area, however, the poles had to be secured from ever greater distances, generally at increased prices. In the peak years of hop production in New York, for example, poles were brought from the northern Adirondacks and Canada by rail, canal, and wagon. In the Northwest, once covered by forests, growers also found it necessary to go farther afield.

The poles were set into the ground as soon as the frost left and it was soft enough to work; an iron bar or auger served to drive the initial hole. They were set deep enough to securely support the weight that would be placed upon them, and earth was rammed around their bases to buttress them against the winds. If two poles were used, they were set about a foot apart, with their tops flaring away from one another, to prevent the tops of the vines from growing together.[14]

Training the crop began with the selection of the proper shoots, each about two feet tall, which became the short and stocky young vines. These shoots were tied to the wooden pole in such a manner that they grew around the vine with the sun.

*2.4  Training the young hop vines on strings, Yakima Valley, 1950. Note that two vines are chosen, and each is trained to follow its string in a clockwise rotation. (Washington State Historical Society, Tacoma, Wash.)*

*2.5  Puyallup Hop Company receipt, May 2, 1895. Initially, training the vines was not considered particulary demanding work. F. More worked for a day and a half training vines, making a total of seventy-five cents. (Meeker Collection, Washington State Historical Society, Tacoma, Wash.)*

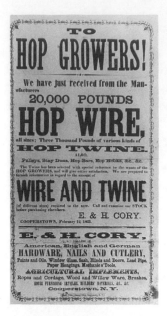

*2.6 Hardware store broadside, Cooperstown, New York, 1863. Wire and twine were available alongside other agricultural implements and housewares. (Courtesy of John Alden Haight.)*

Cotton yarn was preferred for this purpose. The grower and his assistants sometimes used an old stocking, unraveling it as they made their way through the fields.[15] The process had to be repeated as many times as necessary to keep the vine on the pole. If left untended, hop foliage became dense and prevented sunlight from reaching and ripening the cones. Often, before or after a summer storm or a high wind, the vines needed to be secured.

The question of how many vines should be allowed to grow from the root was continually debated throughout the hop-growing regions of the country. In the earliest fields four vines trained to follow two poles was thought proper, although three poles were sometimes believed to be better, with two or three vines apiece. On the other hand, three vines might be allowed if only one pole was used, the argument being that three vines with larger leaves and fruit would be more desirable than four with a smaller cone. For the most part, the third was thought of as being "in reserve," and was later removed if it was not wanted. Stocky and close-jointed vines were the best, for they were the most productive.[16]

The practice of tying the vine so that it would climb up on long "bean poles" came under question in the mid-nineteenth century, as concerns for increased production began to receive ever more attention. One obvious alternative to growing the vine vertically was to allow it to grow horizontally, and perhaps permit more sun to penetrate the vine. One of the earliest and the most widely accepted method of advancing this idea in some regions was the horizontal system, patented in 1863 by F. W. Collins of Rochester and H. W. Pratt of Guilford, New York.[17] This system made use of only one pole, seven to eight feet high, on each hill. The tops of these poles were connected with tarred wool or broom-makers' twine, or wire, and the cords or wires crossed the field at right angles.[18] This system had, it was claimed, four principal advantages. First, the horizontal system allowed the vine to be trained lower, and picking could take place without cutting the vine. The sap could return to the roots, and the dead tops could be removed during the winter. This supposedly allowed the hop to become more vigorous each year. Second, this system had the advantage of reducing the possibility of damage from having the hops pole whipped about by high winds or storms.[19] Third, the grower saved money and time by using these much smaller poles, because they were

TINGED WITH GOLD

easier to set and remove. Fourth, it was easier for sprayers to spread their material on top of the hops than to attempt to reach through them.

Although a few detractors attempted to trivialize the method,[20] growers in Cobleskill, Fayetteville, Guilford, Morris, Hamilton, Pitcher, Rochester, Sharon, Starkville, and Warnersville, New York, all provided glowing testimonials or were referred to as having "model yards" employing the Collins and Pratt system. Promotion in the agricultural press in the Midwest spread the name and use of the patent. News from England also indicated that "the American horizontal system of growing hops" was tried with considerable success in Kent, although it was not quickly adopted there.[21]

Other poling systems were put forward. L. D. Snook of Barrington, Yates County, New York, invented a method of horizontal hop training that involved stakes three inches in diameter projecting up two and a half feet from the ground, with a one-inch hole bored three inches deep in the upper ends. Poles an inch in diameter and four and a half feet long were set into these holes, and horizontal poles of about the same dimensions were stapled to the vertical members to provide a light framework. The purported advantages included the ability to make use of shorter pieces of timber and sticking, greater exposure to the sun for better ripening, ease of picking, and ease of repair and disassembly.[22]

*2.7 The Horizontal Hop Yard, patented by Collins and Pratt, 1863. Although the invention was developed in New York, the most extensive use of this trellis of wire and twine occurred in the hop-growing states on the Pacific Coast. (From F. W. Collins, "Culture of Hops," Cultivator & Country Gentleman 31, no. 795 [April 1868]: 263.)*

*2.8 The pole system of L. D. Snook, c. 1868. The attempt to make use of shorter and less expensive poles was not practical. Snook's system emphasized horizontal growing, with three-inch-diameter stakes holding replaceable one-inch poles. (From L. D. Snook, "Snook's System of Hop-Training,"* Cultivator & Country Gentleman *31, no. 785 [January 1868]: 74.)*

Critics of both the Collins and Pratt and the Snook systems pointed to the fact that the vines were not given sufficient growing space, and that it was difficult to cultivate the roots under such close restrictions.[23] Further, if a wire broke, the damage was even greater than with the old-fashioned pole system. In spite of these criticisms, various horizontal wire systems were used throughout the hop-growing regions of New York and Wisconsin[24] and eventually became the standard in the western hop industry. By the early 1870s some new yards in California were employing low poles with the plants hung from cords or strings.[25] Daniel Flint, the prominent California hop grower, recounted the three ways in which western farmers trained hops. The first method used poles sixteen to eighteen feet high, two to a hill. The second employed redwood stakes eight feet long, one to a hill, with strings or wires drawn horizontally across the field at right angles to each other, fastened at the top of each stake by a staple. The third method was the trellis system, which was "coming more into use than any other, especially where large crops are grown."[26]

Of the three types, the first was the most traditional method. It soon lost favor. The advantages of the short-stake system were, for a time, more obvious than those of the other two methods. In 1892, for example, Ezra Meeker surveyed the country and found that short poles were being used in New York, Oregon, and the newer fields of the Yakima Valley. As a result, the following year he introduced them in western Washington, reducing the height of his fields from sixteen or seventeen feet to only seven feet.[27]

TINGED WITH GOLD

The trellis system, or more properly the high-wire trellis system, soon came to be preferred as the least expensive.[28] Posts six inches square and twenty feet long were set at a forty-five-degree angle as anchors, and posts six by four inches and twenty feet long were set vertically at every sixth hill. Number 3 or 4 wire ran across the posts, and number 6 wire crossed above every row to provide the grid. In this case, two strings of cotton twine were used to train the crop from the ground upward.

California growers favored the trellis system and were responsible for introducing it in a number of locations. In fact, because the Horst brothers of California, who were expanding their hop fields near Independence, Oregon, were the first to use the trellis work on any extensive scale in the Northwest, it

*2.9 Men trimming and training, Puyallup Valley, Washington, June 1911. As the vines grew, the task of training became more time-consuming. The low-wire trellis system required little field machinery. (Washington State Historical Society, Tacoma, Wash.)*

2.10 *The "California System" at Gurley's Ranch, Washington, September 1906. The high-wire trellis system favored by California growers was quickly adopted throughout the Northwest. (Washington State Historical Society, Tacoma, Wash.)*

came to be known as the "California system."[29] With the introduction of hop-picking machines, wire work sixteen to eighteen feet high became the norm. However, the machinery was designed to fit the existing trellis systems, not the reverse, for the distance between rows remained unchanged.[30]

As the harvest season came closer, every development in the weather and every news item regarding the markets became more important. "Hop raising is very much like horse racing; growers watch more closely the progress on the home stretch; and past experience also proves that the raiser is never sure of a crop till, like the winning horse, it has passed under the line."[31]

A violent late-summer storm that would uproot poles and tangle wires, an influx of insects that would weaken the plants, and the spread of a blight or "rust" were specters that haunted the farmer as the crop ripened. The insects most feared were the spider mites (*Tetranychus* spp.) and hop aphids (*Phorodon humuli*). The former drew on the sap of the plant, while the latter excreted a "honeydew" that accelerated mold. Hop mildew (*Sphaerotheca humuli*), sometimes called "blue mold" or "powdery mold," was the most troublesome disease. A serious epidemic broke out in New York in 1909, speeding the decline of hop culture in that state.[32] In many hop yards the loss was complete. The West Coast states seemed to be immune because the high summer temperatures and low humidity proved unfavorable to the disease. The mold did appear, however, chiefly where humidity was higher and the temperature lower, in the Puyallup and Willamette valleys. Western Washington began to suffer in 1929, Oregon followed in 1930, and California in 1932.[33] In many areas, where growers allowed the disease to go unchecked, it threatened the continued production of the crop.

From late May through the summer, most farmers dusted, washed, or sprayed their crops with powders and solutions to try to control these problems. Sulphur and soap were perhaps the most common, but a wide variety of concoctions were employed, including tobacco juice, whale oil and quassia, and carbolic compounds.[34] Chemicals such as DDT and phosphorous compounds were introduced after World War II;[35] more sophisticated insecticides and fungicides are used today. Various sprayers and dusting devices were patented, some hand held and others pulled by laborers, by horses, and later by tractors.

2.11  *Sulphur sprayer in operation, c. 1910. Dry sulphur preparations were preferred whenever tall poles and wires made spraying impractical. (F. M. Blodgett,* Hop Mildew *[Ithaca, N.Y.: Cornell University, 1913], 296.)*

# Here is the Sprayer You Need for Hops . . .

## It's All Metal From End to End

More rugged — more durable — more dependable — and easier to handle than any spray outfit you have ever seen before.

This "Full-Armored" outfit is all metal, including a leak-less All Metal Tank, proofed against corrosion.

Moreover it is shorter and narrower than previous outfits and has a lower center of gravity. Increased pressure and capacity, too, with economy you would hardly believe possible.

It carries the famous BEAN All-Enclosed Royal Pump, equipped throughout with frictionless bearings and automatic lubrication.

Built complete as shown or with power takeoff for your tractor. See your nearest BEAN Dealer or write direct to us for the complete BEAN Catalog. JOHN BEAN MFG. CO., Division Food Machinery Corporation, W. Julian St. San Jose, Calif., 800 S. E. Hawthorne Blvd., Portland, Ore.

# BEAN "Full-Armored" ROYAL SPRAYERS

2.12  *Advertisement for a "Full-Armored" Royal sprayer, John Bean Manufacturing Company, Oregon and California, 1935. This all-metal machine could be built either self-powered or with power take-off for the tractor. (From* Pacific Hop Grower *5, no. 1 [May 1935]: 8.)*

In the inland hop districts of the West Coast, the problem was often simply a lack of water. The necessity for irrigation restricted the hop-growing district in the Sacramento Valley to the river bottoms and land nearby, a pattern still evident with other crops. In the Yakima Valley, where temperatures can exceed 100 degrees Fahrenheit, artificial irrigation came to include not only ploughing furrows to allow water to soak in, but also the use of extensive sprinkler systems, a costly investment for even the largest commercial growers.

The hop harvest began at different times across the country, depending on the variety of hops being raised, the climate, and the growing conditions. Hop picking in New England and New York could be as early as the last week in August but usually was the first three weeks of September. Wisconsin growers harvested their fields at about the same time. In California picking generally began about August 20 and continued for four to six weeks.[36] The warmer climate in the Sacramento Valley meant that ranches there would finish at about the same time that Mendocino and Sonoma County fields and the farms in Oregon began to call for pickers, giving the former an advantage on the export market. In Oregon the southern counties began about a week earlier than those in the north. Picking in the Washington Territory occurred later, generally beginning about September 20 and lasting about a month.[37]

Some judgment about the ripeness of a hop crop could be made when examining the seed, for it began to change from a green or pale straw color to a light brown, and to feel firm, although easily rubbed to pieces. The hops also emitted a "fragrant smell," as the lupulin, or small globules of bright yellow resin, formed in the head of the hop, at the base of the leaves. The lupulin was the only valuable part, for it contained the chemicals that flavored the beer.

When the hops appeared ripe, pickers gathered them with the greatest speed. Hops had to be picked before the first frost, which would injure them greatly. Field hands discovered that a number of green hops and a few brown, discolored hops could be found in the clusters of mature, light brown cones. A few young cones might be admitted, but those that were old or damaged were avoided because they would seriously affect the value of the baled product.

The process of picking began with cutting the vines, generally a foot or two above the ground, and removing the laden

pole to a cleared area immediately adjacent. This was the task of the "tender," who extracted the pole using a device which hung from his shoulders. In the late nineteenth century, as trellis systems became more commonplace, the tender was replaced by a "wireman," whose job of cutting down the vine required a knifelike edge fastened to the end of a long rod.

In the traditional process the poles were laid over long, narrow boxes, so that the hops could be removed. Alternatively, the poles could be placed on a "lug," a pole running the length of the box, elevated about two and a half feet above it, and supported by uprights nailed at each end. In the fields along the Pacific Coast, simple forks were sometimes used.[38] Trellis systems, on the other hand, required no such additional field implements.

As late as the 1830s no standard method of picking hops had been adopted. "The most convenient way," noted one account, was to gather them "into a long square frame of wood called a bin, with a cloth hanging on tenter-hooks within it."[39] Boxes approximately six feet long, twenty inches deep, and about twenty inches wide, made of half-inch boards, as light as possible, were required. The box was essentially a frame for the cloth hanging inside it. Two-inch-thick strips of board, each about ten feet long, nailed to the top of the box and extending out on either side of the box two feet, served as handles.[40] About the middle of the nineteenth century, this traditional box was divided by boards into four equal sections. The cloth used

*2.14  Standard New York hop box with awning, c. 1880. The first grower to divide the box into four compartments was Morris Terry, of Waterville, New York. (From E[zra] Meeker,* Hop Culture *[Puyallup, Washington Territory, 1883], 89.)*

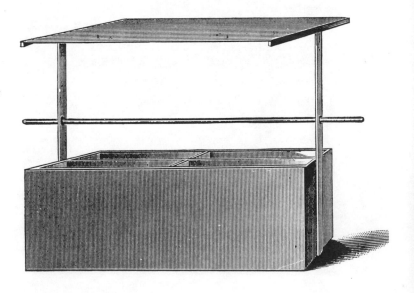

TINGED WITH GOLD

to line the boxes, to carry the hops from the field to the kiln, and to bale the cured cones, was rather special because it had to be wider than usual. In the early nineteenth century, forty-inch-wide English cloth, from Farnham, was preferred in Great Britain and greatly in demand in the United States.[41]

The size of the hop box and the hop sack was a topic of considerable discussion, largely because these containers were used as a measure of the pickers' work. In 1876 the New York State legislature passed a bill to regulate the dimensions of the box. The law mandated a box with an internal space three feet in length, eighteen inches in width, and twenty-six inches in depth. The men carried four of these, joined into one large wooden box, into the fields. The law required that an inspector measure each box, and the name and title of the inspector was painted or branded on the box to ensure accountability. The law further stipulated that the justices of the peace in the towns across the state were to act as inspectors, and it prescribed a number of fees and penalties for violators.[42]

The standard-size hop box in Wisconsin was slightly shallower: thirty-six inches long, eighteen inches wide, and twenty-four inches deep. Four such containers joined together made a quadruple box six feet three inches long and three feet three inches wide.[43] The local hop growers' associations appointed the inspectors and weighers. In Juneau County, for example, three inspectors were appointed to act as a commission, setting the grade of the hops, and two weighers served each of the principal towns in the hop district.[44]

In California the call for a uniform size led to a container seven feet long, three and a half feet wide, and three and a half feet deep; the box was divided in the center.[45] This version, longer, wider, and deeper than those used in New York and Wisconsin, was perhaps a result of the emphasis on commercial growing. In the Pacific Northwest, by contrast, there never was an agreed-upon size. The only requirement was that all the boxes used in a given hop-growing region be of uniform size. At least one reference noted that the "Washington box is a little over twice as large as the Oregon box."[46] Meeker indicated that the standard hop box in the Puyallup Valley was five feet ten inches long by two feet ten inches wide at the top, and four feet four inches in width at the bottom, so designed that it could be carried by the field worker without striking his heels.[47] However, more widespread evidence suggests that the early growers allowed pickers to use any container they pleased. This avoided the necessity of making a permanent investment in boxes, which would have been enormous given the size of some of the fields. Indeed, M. J. Meeker invented a "handy hop bag" of canvas in 1894, which came to be generally used in the valley in the early twentieth century.[48]

The emphasis on boxes lessened as field scales came into commonplace use. In the Northeast, because the bags for gathering generally held only one box, they were light enough to be readily thrown onto the collection wagon in the field. The hop bags in Wisconsin often held two boxes. In the Northwest, if these distinctions existed in the early years, they were not long observed. Bags, boxes, and baskets of every dimension were employed. As late as the 1880s pickers would simply spread an eight-by-ten-foot canvas on the ground and throw the cones in that direction. In Oregon, by the turn of the century, vertical stave baskets became the norm for white pickers, while native Americans often preferred to use their blankets.[49] The ethnic background of the pickers did cause growers to give some thought to how containers were marked. In the Sacramento Valley, each box was numbered in Chinese and English. If a box was rejected at the kiln for being full of "dirty" hops, everyone knew immediately which group was at fault.[50]

Once picked, hops must be dried as soon as possible. Uncured hops left to stand begin to heat, losing their color and altering their chemical composition permanently. At least once a day all the hops that had been picked were gathered for drying. If no kiln was available, the hops were spread thinly to dry, in an airy place, protected from the sun and rain. This commonplace method was used well into the nineteenth century, particularly where only small amounts of the crop had been harvested.[51] If hops were grown in any moderately large quantity, a simple kiln would be constructed. An early kiln in New England and New York would be a small stone structure, banked into the side of a hill. After the middle of the nineteenth century, frame kilns were much more commonplace. These were favored because they were easier to construct and required less capital investment.[52] Early kilns had open fires, above which a slatted bed of hops was suspended, but the open fire was soon replaced by wood-fired cast-iron stoves.[53] Later, furnaces were devised, and a number of air-blast varieties were developed to force air through the hops. In the mid-1930s,

2.15 *Hop house, Wood-stock, Maine, c. 1880(?). This rare frame survivor is banked, in the manner of many hop houses in New England. (Kirk Mohney, Maine Historic Preservation Commission.)*

2.16 *Advertisement for "The Granger" stove, H. D. Babcock, Manufacturer, Leonardsville, Madison County, New York, 1878. Several foundries in the Waterville area were set to work meeting the need for hop stoves. (From* Waterville Times, *September 26, 1878.)*

2.17 *Advertisement for a cast iron corrugated hop furnace, Eugene Foundry & Machine Co., Eugene, Oregon, 1935. The corrugated surface increased the amount of surface radiation. (From* Pacific Hop Grower 5, *no. 1 [May 1935]: 3.)*

although cordwood remained the most commonly used fuel, new installations employed oil furnaces that generated steam, the heat from which was blown through the crop. Large commercial operators favored this method.[54] Oil-fired systems, in turn, were superseded by gas-fired furnaces wherever economics dictated.

The grower with limited experience, however, often regarded building a kiln as an experiment. If the kiln proved successful, it would be made available to neighbors.[55] For the grower who did not have the capital to invest in a kiln, it was common to take the crop to a nearby kiln and to pay for the service of drying. In New York in 1865, for example, the cost of drying was about two cents per pound, and the dryer furnished

*2.18  Gas-fired furnace, Jackson Road, Sacramento, California, installed c. 1950(?). The tremendous amount of heat needed for drying caused growers to adapt their kilns to cheaper fuels.*

the kiln fuel and sulphur for bleaching the hops, if needed. In Wisconsin, only a few years later, two and a half to three cents per pound were paid for this service.[56] Further, the grower might also pay for the labor of baling and for sacking cloth.

The process of loading the kiln did not vary greatly from one region to another. Generally, the operator heated the kiln before the first load of hops was placed on the drying floor. The hops, carried in boxes, bags, or large bushels from the field, were raised by means of a ramp, steps, or block and tackle to the upper floor of the kiln, and spread on the drying cloth. In some instances the hops were dumped in the center of the hop kiln floor and moved by hand to the edges; in other cases barley forks with tines turned slightly upward were used to carefully, loosely, and evenly distribute the fragile cones.[57]

The depth of the hops on the floor depended upon the nature of the kiln or dryer. In most kilns built during the first three quarters of the nineteenth century, hops were laid to a depth variously prescribed as between eight and twelve inches. Many existing kilns have horizontal chalk marks on the walls above the drying floor, indicating the level to which hops were loaded, with one or more lines perhaps four or five inches lower, to indicate the level to which the hops would settle when completely dry. Some kilns on the Pacific Coast display permanently scribed lines at regular intervals on the entire wall surface, indicating that more varied drying techniques were used. In these instances, hops would be laid from twenty to thirty inches deep, although it required from sixteen to twenty hours to dry each load.[58] With the introduction of fan-blast kilns and vacuum processing, hops could be laid deeper; hence the more numerous markings.

The kiln, closed on all sides at the second-story level, was charged from below with sufficient fuel, and a fire or fires were kindled either on the ground floor or in a furnace at that level. With the continuous application of heat, the water vapor in the hops became steam, which was driven up. During the nineteenth century, the best drying temperatures were thought to be in the range of 140 to 160 degrees Fahrenheit, but later experimentation led to the use of a temperature between 100 and 140 degrees, with a longer drying period, depending upon the volume of air being pushed through the crop, the moisture content of the hops, and the relative humidity of the air.[59] The

TINGED WITH GOLD

operator used a thermometer hung over the middle of the hop floor to monitor the temperature.

If the hops were at all discolored, a small amount of sulphur might be burnt under them to impart a more uniform appearance. The recipes varied, from a few tablespoons to twenty-five or thirty pounds per hundred boxes of hops. Other chemicals were also tried, but growers who had a nearly flawless crop dispensed with this practice, and numerous reference made clear throughout the hop-growing regions of the country that bleaching brought no real benefits.[60]

When sulphur was used, it was burned as soon as the hops were completely warm and moist. After the fumes filled the kiln, the operator opened the upper windows, vents, or cowl, and a strong draft of air was let in from the bottom to force the hot air through the hops and drive the sulphur and the hot steam out of the building. One of the most important factors was the direction of the prevailing winds. Unexpected drafts of cold air on the hops at this point were dangerous; any condensation which fell back on the crop could discolor it.

After all the moisture was driven off, with the hops on the top feeling a little dry, most kiln operators thought it proper to turn the load, with the heat slackened a little. Other operators believed it just as important not to turn the hops, for fear of breaking the cones. One reason for turning was to increase the evenness of drying, for some cones lay closer to the source of heat than others, and some might have been packed more densely than others. In cases where the heat rose through areas of lightly packed hops, the air rushed through and dried the cones quickly, diverting the warmth from other areas more in need. Turning obviated this difficulty, for all the cones dried equally. Stirring the hops was accomplished in various ways: turning with large, light, square scoop shovels, stirring with long-handled rakes, and even stirring with the feet. Although the first two were most commonplace, according to one writer, the method most in use in the Wisconsin hop region was called "plowing with the feet." "The person keeps his feet close to the floor, so as to stir the bottom, and not to tread on the hops, and, by short steps, goes back and forth across the room, (or around as a ploughman goes 'round his 'land') making the furrows near enough to thoroughly stir the hops."[61]

Drying hops was decidedly the most critical operation. Curing too little would mean the hops would discolor and lose flavor, while curing too long made them too brittle. Any one of several tests could be used to determine whether the load was sufficiently done. "A very simple one and perhaps as often used as any, is to examine the stems, and when about one-third of them are dry and brittle, the others being more or less wilted, they may be considered about done."[62]

To remove the hops from the kiln, the operators used long-handled "scuppers," shovels designed to move the fragile cones with a minimum of breakage. A number of late-nineteenth-century inventors attempted to develop movable trays and conveyor belts to aid this process, and many schemes were adopted, particularly in the West. For example, by the late 1880s Meeker built his large commercial operations in Puyallup, Washington, with the cooling barns almost completely divorced from the kilns. They were connected only by elevated

*2.19  Long-handled scupper and hop-pole-pulling harness, Oneida County, New York. Special scoops were designed to avoid breaking the fragile cones. (Courtesy of John Alden Haight.)*

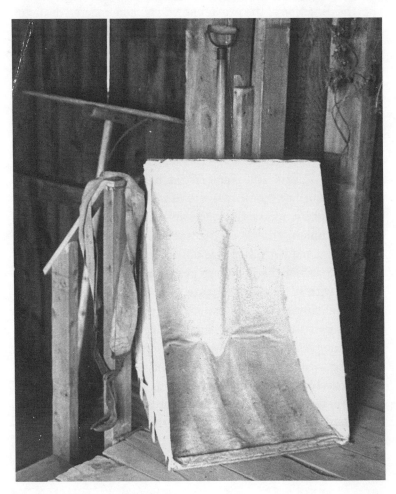

TINGED WITH GOLD

tracks for the hop cars. This allowed the cooling barn to be enlarged to accommodate hundreds of bales.

At this stage in the process the hops were set aside to regain gradually some of the moisture that they had lost, so that they would not be broken when they were bagged. The principle was well understood. In 1801, for example, the instructions for bagging stipulated that growers should protect newly cured hops from air for three or four days by covering them with a blanket. "When the hops are so moist that they may be pressed together without breaking, they are fit for bagging."[63]

The earliest methods of pressing the crop were similar to those used in Europe. The process required coarse linen bags, usually about eleven feet long and seven feet in circumference, although any size could be used. First a hole was cut through the floor, large enough for a man to pass through. Next the open end of the bag was fastened to a hoop, larger than the hole. The bag, inserted into the hole, was supported by the floor below. As the dried cones were shoveled into the hole, they were regularly pressed down by a man lowered into the bag, until it was full. Finally, the bag was sewn shut, labeled, and stored or shipped to market.

The idea that hops had to be trodden down into the bag by the force of human weight remained commonplace through the 1830s in the United States and through the 1880s in Great Britain.[64] It was an unpleasant, dusty task, advanced somewhat by using a board on top of the hops in the bag, tamping with a springy motion of the knees. This method was preferable because it created less dust and fewer broken cones.

The bagging process could be advanced even further by the use of a press. Screw presses were introduced in Massachusetts from England in 1797 or 1798.[65] However, the English practice of pressing in cylindrical "pokes" seems to have been almost completely eschewed in the United States soon after commercial production began, because early-nineteenth-century shipping notices refer to bales. The presses were permanent devices, fixed in place: wooden screws turned in a hole tapped through one of the second-story floor joists of the barn.[66]

Screw presses became independent of the structure by the late 1840s. This change affected the amount of sacking needed and altered the method of stitching and sewing. Presses designed to produce squared bales consisted of two wooden up-

2.20 *Draft and fan-blast kilns, Ezra Meeker & Co., Puyallup, Washington Territory, c. 1888. One of the most elaborate layouts in the Pacific Northwest, the complex grew to include ten kilns. The water barrels on the roof were an ever-present reminder of the danger of fire. (Meeker Collection, Washington State Historical Society, Tacoma, Wash.)*

rights with a beam across the top and two sills at the bottom to support the posts and secure them on the floor. A screw passing through the center of the beam went through a box made of planks, constructed so as to be easily disassembled. Such a box would be about four and a half feet long, one and a half feet wide, and four to five feet high. When the box was ready, a cloth was placed inside it, and hops were poured in until it was full, when another cloth was put on top. A flat board, termed a "follower," was then put on, and the screw turned down as far as necessary, after which the sides of the box were removed and the sides of the cloth sewn. The screw was turned up, the bale of hops taken up, and another piece of cloth sewn to each end. Last, the bale was weighed, marked, and made ready for market.[67]

Stencils were generally used to mark one or more sides of the bale with the name or initials of the grower. The name of the consignee was often stenciled on the top. Turpentine and lampblack were preferable, for water-based paints were impermanent and oil-based paints were slow-drying and injured the contents.[68]

"Hand presses" were commonplace throughout the hop-growing regions of the country. In places where the quantity of hops being produced warranted a larger investment, horse-powered presses were substituted, but these were comparatively rare. Although English hop growers had experimented with hydraulic presses by 1838, these do not seem to have been available on this side of the Atlantic.[69]

Portable presses, so called because they could be moved

*2.21 The Harris press, Waterville, New York, complete* (left) *and with front removed* (right), *c. 1860. This was the first so-called portable press, widely recognized as inexpensive and easy to operate. (From E[zra] Meeker,* Hop Culture *[Puyallup, Washington Territory, 1883], 86.)*

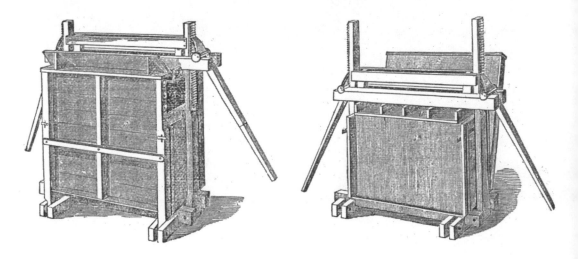

TINGED WITH GOLD

inside the room in which they were placed, became very popular during the Civil War era and superseded the screw press within a few years. Perhaps the most famous portable press was that invented by L. W. Harris in 1859.[70] A device made of wood with movable cast-iron parts and reinforcing, it used a ratcheting device rather than a screw. Most important, the Harris press cost only about fifty dollars. When announcing that Harris had received his patent in August 1860, the Waterville, New York, paper wished him well: "May he become as rich as Sir Richard Arkwright, of spinning jenny notoriety, who, starting life a poor barber-boy, died worth 500,000 pounds sterling."[71] The demand quickly exceeded the supply. Within a few years, other manufacturers were making and selling the Harris press, not only in upstate New York but also elsewhere in the country. Similar patented devices were manufactured and sold in most of the major hop-marketing towns. In the Pacific Northwest, for example, Puyallup provided presses for growers in Washington and Oregon. These were generally of wooden construction with wrought and cast iron parts.[72]

2.22  *Wiscarson hop yard, Santa Clara, Oregon, c. 1895. The press required a team of men to bale the hops properly and move the full sacks. (Lane County Historical Museum, Eugene, Oreg.)*

Inevitably, situations arose in which even a number of portable presses would prove unequal to the job, requiring further mechanization. On the large eastern farm or western ranch, growers built horse-powered presses using rack-and-pinion principles. Steam-powered presses seem to have been first installed in warehouses—B. A. Beardsley built one in the Putnam Street Warehouse in Waterville in March 1868—and they were undoubtedly available to the enterprising grower soon afterward.[73]

All of these inventions, however, led to pressing too many hops in a bale. The Meeker Company in Puyallup, for example, served notice that they would not accept bales weighing over 210 pounds, for they would not sell. An average of 170 or 180 pounds was more desirable because the hops thus packed were thought to retain their freshness longer and fetch a higher price at market. One early-twentieth-century source believed 185 pounds was the optimum.[74]

Although various mechanical balers were used until the mid-twentieth century, with the advent of rural electrification the use of the electric baler became commonplace. These were

*2.23 Automatic electric baler, Sloper Plow Works, Independence, Oregon, 1935. The introduction of electrical power in rural areas greatly lessened the amount of physical labor and shortened the time involved in baling. (From* Pacific Hop Grower *5, no. 2 [June 1935]: 4.)*

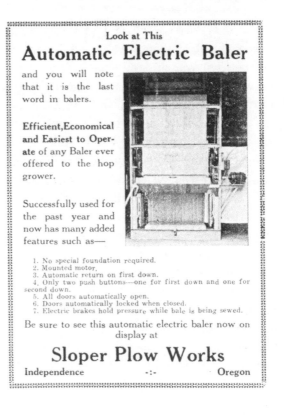

Look at This
## Automatic Electric Baler

and you will note that it is the last word in balers.

**Efficient, Economical and Easiest to Operate** of any Baler ever offered to the hop grower.

Successfully used for the past year and now has many added features such as—

1. No special foundation required.
2. Mounted motor.
3. Automatic return on first down.
4. Only two push buttons—one for first down and one for second down.
5. All doors automatically open.
6. Doors automatically locked when closed.
7. Electric brakes hold pressure while bale is being sewed.

Be sure to see this automatic electric baler now on display at

## Sloper Plow Works
Independence        -:-        Oregon

considerably easier to operate and maintain. More recently, hand-held electric stitching machines have been used to seal the burlap bags.

As has been shown, a tremendous amount of effort went into preparing the crop for shipment. Storing the bales was considerably simpler: the bales were supposed to be set on end, relying upon one another as little as possible for support, in a cool and dry environment. Whether the farmer stored the bales in his hop house or barns, or whether the dealer kept the baled crop in his warehouse on a railroad siding, the atmospheric conditions needed were the same.

The idea that the industry could be rid of the problem and expense associated with baling and storing hops surfaced repeatedly during the early and mid-nineteenth century in New York State, as chemists began to experiment with the process of extracting the lupulin and leaving behind the flower. As one editorial of 1851 noted, the cost of separating and packing the lupulin would be considerably smaller than the expense associated with the handling and transportation of "the useless leaves in addition to the loss of malt liquor soaked up by these leaves and lost both to the manufacturer and the consumer."[75]

*2.24  Warehouse, West Yakima, Washington, October 1908. Hops and apples were stored side-by-side in this storehouse constructed of local stone. (Washington State Historical Society, Tacoma, Wash.)*

2.25  *New York Extract Company works, Waterville, New York, c. 1900(?). The idea of shipping only the essential oils to market led to this first major commercial effort at production.*

The knowledge that such a process was possible had come from an understanding of chemistry in medical research. A Dr. Ives of New York City was the first person to isolate the resinous globules, which he called lupulin. His findings were first published in 1835. He found that the substance forms a few days before the hop ripens. French and English chemists confirmed that nothing but the lupulin was of value, for it alone contained the bitterness and aromatic flavor of the hop, which was considered essential to the excellence and preservation of malt liquor.[76]

With further experimentation, three decades later, the extraction process became commercially viable. In 1865 a patent was granted to Professor Samuel R. Percy and W. S. Wells of New York City for a method of removing lupulin. Dr. Percy, a chemist by training, conducted experiments by steeping hops in water solutions until all their value was dissolved. He then mixed this liquid with barley sugar and glucose and evaporated the water in a vacuum pan, in a process very similar to that by which Borden's condensed milk was made. When the condensation was complete, a semifluid extract with the viscosity of honey was created, and this was kept in air-tight containers until needed.[77]

The patent was put to use in 1867, when a Mr. Hawks, in Rochester, New York, built a factory for condensing the extraction of both malt and hops. Another patent, awarded about the same time to W. A. Lawrence of Waterville, New York, was used by the New York Extract Company at their works in Long Island City and in Waterville. The company received nationwide attention at the Centennial Exhibition in Philadelphia, for

TINGED WITH GOLD

the extract was awarded a gold medal, and beer and ale made with the extract were also given premiums.[78] In May 1877 the Waterville factory was consuming about fifty bales of hops a week. By February 1878 it was running two shifts and consuming twenty-five bales of hops per day.[79] Producing the extract on a year-round basis, the company received and quickly filled orders from all over the world. About 1883 a licensee, J. W. Whiting, erected another factory in Waterville, which had a capacity for extracting a hundred bales of hops per day.[80] The idea proved to be a good one, except that the business depended upon a local supply of hops. Hence, the decline of this company followed the general decline of hop growing in New York State, as other extract companies were begun and prospered on the Pacific Coast. The work of razing the Waterville buildings of the New York Extract Company was begun in the summer of 1935.[81]

From one side of the country to the other, hop culture involved certain tasks, which were undertaken at specific times during the year. For those who were familiar with processes involved, the landscape was immediately understandable. The stacks or grid of poles were recognizable in winter, long before the winding vines were first trained in spring. And the tall verdant growth with its clusters of cones was easily distinguished in the summer, before the smells of the kilns and dryers at harvest season. These were the keys, expressions of a culture that had a common verbal vocabulary, understood by all who knew something about the crop. Although many persons in allied walks of life, notably commerce and transportation, could claim some knowledge of hop culture, the two principal classes most deeply involved with hop culture were the growers and the pickers.

# ❧ 3 ❧
# The Grower's Perspective

From the grower's perspective, the most important reason for raising hops was that the crop offered a chance to make a significant amount of money relatively quickly. The glowing reports of growers, carried to an ever-wider audience by the expanding agricultural press, spurred an increasing number of farmers to try the crop. Money provided the impetus, and growers saw any problem as both a challenge and a gamble. The three principal concerns of the hop grower were the amount of start-up capital that was required, the vicissitudes of the marketplace, and the problems associated with the army of pickers needed at harvest time. The grower's initial investment was a large one, an act of faith and confidence. If only the prices would not fluctuate and the middlemen would cooperate; if only the pickers would perform as flawlessly as the ideal picking machine, the grower might be happy with his profits.

Starting a hop yard was an expensive undertaking. The upfront costs included preparing the soil; manuring, cultivating, and laying out the fields; setting out the roots; procuring poles, wires, twine, and burlap; obtaining dusting and baling equipment, picking baskets, boxes, and sacks; and building a kiln. A brief look at the expenses of a few small growers in central New York further emphasizes the point that harvesting hops was a very labor-intensive, expensive task.

James H. Dunbar of Hamilton, New York, kept meticulous records of four successive crops grown on four acres of his farm. In 1837 he recorded the following expenses and income:

To sticking poles, cultivation and tending through the
    season, . . . . . . . . . . . . . . . . . . . . . . . . . $50.00
To picking, at twelve and one-half cents per box, . . . . $106.75
To box tending, including board . . . . . . . . . . . . . 45.37
To boarding girls, while picking . . . . . . . . . . . . . 65.00
To 110 yards hemp bagging at 20c. per yard . . . . . . . 22.00
To pressing 8,060 lbs. hops, at twelve and one-half cents per
    hundred . . . . . . . . . . . . . . . . . . . . . . . . 10.07
To coal and labor in drying . . . . . . . . . . . . . . . 49.50
To interest on first cost of land, at $60.00 per acre . . . . . 16.80
To transportation to Utica . . . . . . . . . . . . . . . 12.00
To decay of poles . . . . . . . . . . . . . . . . . . . . 6.00
      Whole expense . . . . . . . . . . . . . . . . . . $383.99

8,060 lbs. hops, sold on contract at 14c. though market was but
    5c. amounting to . . . . . . . . . . . . . . . . . . $1,128.40
Deduct expense, leaves a gain of . . . . . . . . . . . . 745.41

Dunbar's profit was impressive, but it depended upon an extraordinarily high yield—just over two thousand pounds per acre—and an above-average price—fourteen cents per pound. The profit indicated is slightly higher than it should be, however, because the account is not complete. Dunbar did not include the costs of manure or poles, as the yard was ready for working when he purchased the farm. He also omitted the cost of a kiln that was standing on the property.[1] Even with these oversights, the cost of labor is more than the sum needed for materials.

About 1850 Dunbar sold this farm to H. P. Potter of East Hamilton, Madison County, who reported these figures for the 1851 season:

*Lot No. 8—Hops—Five Acres*                                    *DR.*

To 109 loads of manure and drawing, 4s . . . . . . . . . $54.50
One hundred thirty-four and one-half days of work, to Sept. 1st,
    6s. . . . . . . . . . . . . . . . . . . . . . . . . . . 100.88
42 da. tending box, 6s. . . . . . . . . . . . . . . . . . . 31.50
15 da. drying hops, 12s . . . . . . . . . . . . . . . . . . 22.50
14 da. pressing hops, 6s. . . . . . . . . . . . . . . . . . 10.50
270 bushels, coal, 7c . . . . . . . . . . . . . . . . . . . 18.90
Hop sacking and twine . . . . . . . . . . . . . . . . . . 24.32
Cost of engaging and collecting hop pickers . . . . . . . . 13.18
Paid for picking hops . . . . . . . . . . . . . . . . . . 132.36
Help in house. . . . . . . . . . . . . . . . . . . . . . . 9.00
Provisions . . . . . . . . . . . . . . . . . . . . . . . 48.65
Interest on land, $65.00 per acre. . . . . . . . . . . . . 22.75
Depreciation on hop poles, 10 percent . . . . . . . . . . 30.00
Depreciation on hop-house, 10 percent . . . . . . . . . . 30.00
Insurance . . . . . . . . . . . . . . . . . . . . . . . . 2.75
Whole expense . . . . . . . . . . . . . . . . . . . . . $551.79
Cost per acre. . . . . . . . . . . . . . . . . . . . . . 110.36
Profit of five acres. . . . . . . . . . . . . . . . . . . 1,050.79
Credit . . . . . . . . . . . . . . . . . . . . . . . . $1,602.20

                                                                *CR.*

1851 Sept 20. 7,801 lbs, sold 20c. . . . . . . . . . . . . 1,560.20
Hop roots sold in April . . . . . . . . . . . . . . . . . 42.00
                                                          $1,602.20[2]

Potter's itemized list is more detailed than his predecessor's and thus provides a more accurate reflection of the costs involved. The high yield and a low market price would have

TINGED WITH GOLD

produced a deficit were it not for the sale of hop roots, which allowed the grower to balance his books. The yield in upstate New York was generally admitted to be about six hundred pounds per acre,[3] and the average price for 1851 was thirty-five cents per pound. With these figures Potter would have lost even more. The cost per acre is $110.35, of which $63.80 was paid to hired help.

One more example, from Waterville, New York, indicates the cost of raising one acre of hops yielding one thousand pounds, in 1878:

1556 poles at 11 cents each, $171.16
Interest on same at 7 percent, . . . . . . . . $11.98
Depreciation of poles, 10 percent . . . . . . . 17.12
Taxes, $1.00, fertilizers and cartage . . . . . . 14.90
18 days' work, man or team, cultivating, hoeing
    and grubbing . . . . . . . . . . . . . . $18.00
Picking eighty-three and one-half boxes, 50 cents per
    box . . . . . . . . . . . . . . . . . . $41.67
Tending box . . . . . . . . . . . . . . . . 8.34
Emptying boxes, superintending yard . . . . . 2.25
Teaming . . . . . . . . . . . . . . . . . . 2.50
Dryer and assistant, two kilns per day . . . . . 4.00
Coal, $2.00, brimstone, 30 lbs. at $1.05 . . . . . 3.05
Pressing five bales . . . . . . . . . . . . . 1.25
Depreciation of kiln, cloth and sacks . . . . . . . .80          $63.86
Use of the hop-house, costing $800 . . . . . . . 5.33
Insurance on hop-house and hops, 30 da. . . . . 1.50
35 lbs. sacking, at 8½ cents per lb. . . . . . . . 2.98          $9.81

                                    Total:      $135.77

At a price of ten cents per pound, which was three cents lower than the average for the year, this acre lost $35.77.

Although this is an extremely limited sampling, a review of all three examples indicates that the cost per acre was increasing, from about $96 in 1837, to $110 in 1850 and $136 in 1878. The prices of fertilizer, poles, land, and labor were escalating rapidly. In fact, Meeker noted that from 1878 to 1883 these costs had risen on the average another 50 percent. Although inflation is a factor to be considered, the rising costs were not offset by a parallel rise in prices. This meant that only when the market price was high and the yield well above average could the grower expect to gain a profit. When the market was soft, or if the fields suffered from lack of fertilizer, plant disease, or insects decreasing the yield, the grower suffered a loss.

If the layout was a large one, the grower's concerns multiplied. The quantities of supplies needed for a large commercial operation were staggering. For example, the Meekers found, when planning for their 611 acres in the Puyallup Valley, that they required poles numbering in the tens of thousands, over eight thousand pounds of twine to string their yards, and fifteen hundred gallons of whale oil and twenty-five hundred pounds of caustic soda for spraying.[4]

On large farms or ranches, the need to accommodate the pickers led to an extensive building program. Growers constructed cabins or dormitories, and any number of auxiliary structures. Other preharvest camp expenses included not only providing housing, food, and fuel for the permanent employees but also the costs of ditching and draining fields, and creating roads and bridges. Harvest expenses encompassed tents, camp stoves, barrels for water and garbage, straw for bedding, and medicine. Insurance was extended to include indemnity from fires, crop failure, and forgery.[5] In the nineteenth century the transportation of pickers and the baled product sometimes involved hiring teamsters. In the twentieth century these expenses became associated with automobile and truck operation. In all, the number of people involved and the number of details that needed to be considered made hop growing a bookkeeping headache.

Hop growers, like many other farmers, faced a relatively inelastic demand for their crop. When a small crop was harvested, the grower benefited from a seller's market. When a large crop was produced, the grower often found that a reduction in price was necessary to interest buyers. The relationship between price at market time and the possibility of exporting the crop was also critical, for already by the early nineteenth century, a good crop amounted to about two-fifths more than could be consumed domestically.[6]

One of the greatest challenges that a hop grower faced was anticipating the market. There were times when the crop was excellent and the price was low. In other years, when the harvest was poor, the prices were high. In 1833, for example, one farm on the Connecticut River in Vermont produced three thousand pounds, which sold for twenty cents a pound; the

TINGED WITH GOLD

following year the same farm expanded its operation to five and a half acres, and the yield was four thousand pounds, but this sold at only fifteen cents a pound. In the first year, after all the expenses were tallied, the farm made one hundred dollars, but the second year was a considerable loss despite the increase in yield.[7]

The swing in prices paid for hops continued in succeeding decades, a problem largely attributable to the failure of hop crops abroad. In the years from 1880 to 1910, hops sold for as little as $.03 per pound and for as much as $1.13 a pound. In the period from 1930 to 1950, the price ranged from $.10 to $.69.[8] This unpredictability caused some producers to become so discouraged that they would plow up their yards at exactly the time when they should have been planting more roots.

Even within the same season, the price varied wildly. In 1846, for example, the price on the Boston market began at nine and a half cents, ran up to thirty-five cents, and by the end of the year dipped to twenty cents.[9] July 1 of every year was the beginning of a new season, and prices for hops on hand were adjusted downward. Top prices were generally obtained at the end of October, although the range fluctuated considerably as large quantities were brought to market.

The grower's anxieties upon entering the market were also due to the difficulty of reaching a fair price. Whereas on his farm or ranch the hop grower was king, when he went to market he was but one player in a more complicated business world. There were others better informed about prices and equally adroit at sensing a bargain.

Growers often bragged about their crops before harvest season, believing that this "advertising" would help secure an adequate number of pickers. Smart, experienced field hands shied away from damaged or low-yielding fields, for they could make more money in the better yards. The grower's promotion also had the important side effect of bringing the curious hop buyers into the countryside, for it was difficult to estimate the value of the yield before the harvest. The fields in one area could be adversely affected by weather conditions, blight, and insect infestation, while another would produce a bumper crop.

In the early nineteenth century brewers visited the fields well in advance of the harvest, promising farmers reasonable commissions and moderate storage rates if they would consign the sale of their crops.[10] A well-known brewer like Matthew Vassar would be welcomed in the hop region of central New York as a patron of importance. Vassar was a particular friend of Gurdon Avery and the Palmer family, who were the hop kings in their day.[11]

By the late 1840s the volume of production in a number of areas led some of the major growers to realize that more money could be made in trading hops than in growing them. Several central New York growers became brokers and established offices in centers such as Waterville, Cooperstown, and Schoharie. For example, the father-and-son team of Mortimer L. and Daniel Conger of Waterville, well-known growers, began to deal in hops in 1847. The Congers maintained large, active hop fields in Oneida and Jefferson counties, while buying and selling hops for others. William P. Locke and his two brothers established their dealership in 1864, H. W. Tower and Son began in 1867, and Charles Terry started in 1869.

All of these early dealers traveled widely in the hop-growing regions of the United States. Locke contracted with growers throughout New York and dealt in baled hops, hop roots, sack-

*3.1 Letterhead of Daniel Conger & Son, and the business card and hop sample ticket of William P. Locke, all of Waterville, New York, c. 1880.*

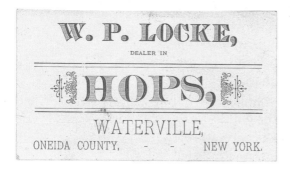

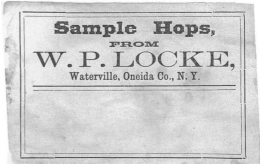

TINGED WITH GOLD

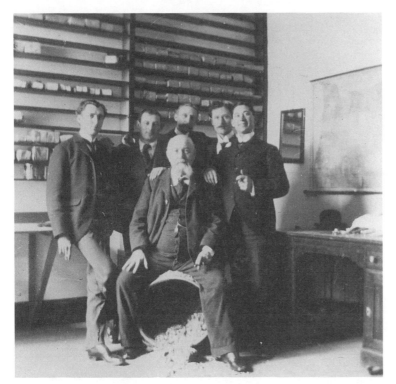

ing, and kiln cloth from the Mississippi Valley to Massachusetts.[12] Terry is known to have visited Kilbourn City, Wisconsin, to purchase hops, competing against G. J. Hansen, the foremost representative of a family of dealers in that area. Terry also competed against Richard Weaver, one of the largest growers in adjacent Waukesha County.[13] At the same time, hop growers from Loganville, Wisconsin, appointed representatives to visit eastern and southern hop markets and to learn about the markets and prices. Other city-based commission dealers merely advertised, anticipating that either the grower or a country dealer would contact them when appropriate.[14]

As should be obvious from the names of the firms above, many dealers were members of closely related families that made their fortunes by sharing information. In the Pacific Northwest F. M. Lempleton, working as an agent for Ezra Meeker and Company, dutifully wrote home at least once a week after collecting commitments and promising shipments. Each letter began "Dear Grand Pa." Competition was so severe that the agents of the Meeker firm wrote in a secret code,

Received by

Fox & Searles.

Sent to E. Meeker & Co

Puyallup

Oct 9th /94

| | | |
|---|---|---|
| Hag. | We accept your offer/ | |
| Backward | 100 Bales | |
| | Will | |
| Labarum | Ship as soon as possible | |
| Tagrag | How shall we draw | |
| Paddlebox | Horst Bros | Oct 10th |
| | offering | Desgobar You must not draw on us |
| Abdomen | 7 ¢ | for |
| | here today | Backward 100 Bales |
| | | must |
| | | Clipso Insist upon fulfilment |
| | | of Contract |
| | Oct 10th | Maltrabaja 25th day of May |
| Serbanya | I cannot ship | Screpare Have you sent |
| Backward | 100 Bales | renewal |
| | Without drawing | Trionfo 1500 |
| Piegatura | 4 ¢ per lb | Desfuia Draft will be due |
| Atspasen | If you do not advance | Obstetriz 16th day of October |
| | we | |
| Schlippe | Will sell as soon as possible | |
| | here and | |
| Revoolsi | and will remit | |
| Retwriron | 2500 | |
| | mailed renewal | |

designed to foil eavesdropping on their telegraph messages. For example, "Hag Backward Labarum Tagrag Paddlebox Abdomen" meant "We accept your offer; 100 bales will be shipped as soon as possible; How shall we draw; Horst Brothers are offering $.07 here today."[15]

In essence, growers began to release the responsibility for getting their hops to market to middlemen, some of whom were from their own ranks. These middlemen developed interests and even a language which was all their own.

The market became even more crowded because, as the beer industry grew, brewers no longer came to the field to make contracts directly with the growers or dealers but sent their representatives. The agents for both domestic and foreign brewers were often found in hop-growing regions of the country, perhaps a month before the harvest, surveying the hops in various sections. Factors, merchants who bought and sold hops in their own name and controlled possession of the crop, were also seen in the fields. In early August 1878, for example, a Mr. Vuylsteke of London visited the Waterville, New York, area, returning to England to make his report before going on to Bavaria, Holland, and Hungary.[16]

After the Civil War, buying, selling, receiving, and shipping became increasingly removed from the control of those who raised hops. Growers regularly pledged their crops to a dealer or brewer's agent before the harvest, sometimes years in advance. Presumably this reduced the grower's risk because the price was fixed regardless of market fluctuations. Equally important, growers who contracted their hops did not need to borrow money at the bank to pay their pickers, because the buyers advanced money in the spring and fall to cover the expenses of cultivating and harvesting.[17]

The practice of commissioning hops flourished, and many prominent growers endorsed the idea. In fact, on the West Coast many fields were begun with only that in mind. In California, almost from the earliest days of commercial production, growers raised many more hops than could be used by local brewers. Commission houses grew by leaps and bounds, aided by favorable freight rates on the transcontinental railroads.[18] In San Francisco, Davis and Sutton of New York City provided much of the early advice and solicited business through their circular, although the local firm of Philip Wolf and Company

*3.3 Telegram notebook, E. Meeker & Company, Puyallup, Washington, October 9, 1894. Because of the need for secrecy, telegrams sent to and received by the Meeker firm were in code. The ledger, printed by the dealers Fox & Searles, is arranged to be read as a dialogue between the two columns. (Meeker Papers, Washington State Historical Society, Tacoma, Wash.)*

Established 1862.

## EMMET WELLS' WEEKLY HOP CIRCULAR

OFFICE AND WAREHOUSE

### No. 69 PEARL STREET,

### NEW YORK CITY.

CONSIGNMENTS RESPECTFULLY SOLICITED.

SUBSCRIPTION:

**$3.00 a Year in Advance.**

POSTAGE FREE.

REMIT BY POST OFFICE MONEY ORDER
OR BY REGISTERED LETTER.

WE MAKE THE HOP BUSINESS
A SPECIALTY,
AND DEAL
ONLY ON COMMISSION,
NEVER
Purchasing, Selling or Speculating
ON OUR OWN ACCOUNT.

VOL. XIX.     New York, Friday, Aug. 26, 1881.     No. 51.

## NEW YORK MARKET.

THE nineteenth year's publication of the CIRCULAR closes with this issue. Excepting the fact that there was no change in the price of choice Hops from 23 cents per lb., during the ten months intervening October 15th and August 15th, it has been rather an uneventful season. The receipts into New York for the season amount to 100,000 bales in round numbers, 43,000 bales of which have been shipped to England; the balance (57,000 bales) have been nearly all taken by our local brewers. Large quantities of Hops go direct to Western brewers from the Cooperstown, Waterville and Utica Markets, on orders received by dealers there and in this city. Home crop advices this week are, on the whole, more favorable. The yield will come but little short of last year's, and the quality still promises to be fine. Picking has already commenced in several districts—too early it would seem —though the late hot and forcing weather may have ripened the Hops sufficiently to warrant it; upon this subject we will be better able to pass judgment when we see the Hops. There has been a large trade with brewers this week at unchanged prices, all holders showing a disposition to clear off the old stock and commence business with clean boards.

### RECEIPTS, EXPORTS and IMPORTS.

| | |
|---|---|
| Receipts for the week. (Officially reported on 'Change,).. | 915 bales. |
| Total Receipts since Sept. 1st, 1880,.... ... ... ...... | 96,988 " |
| Total Receipts for same period in 1879,.... ... ... ...... | 82,608 " |
| Export clearances for the week to Europe, ....... ... ... | 181 " |
| Total Exports since Sept. 1st, 1880, .... ... ... ...... | 43,027 " |
| Total Exports for same period in 1879, ... ... ... ...... | 43,954 " |
| Imports for the week,...... ... ... ........ ... ...... | 16 " |
| Total Imports since Sept. 1st, 1880, ... ... ... ...... | 2,094 " |
| Total Imports for same period in 1879, .... ... ... .... | 2,772 " |

### ENGLAND.

LONDON, Aug. 13.—Crop prospects continue about the same, although it is feared that the yield will not be quite so heavy as we anticipated a week or ten days since. The slack vine this year forms rather a large percentage, and very few have the courage to estimate the coming crop at more than £260,000 old duty. The quality bids fair to be above the average of late years, although there will probably be a sprinkle of mouldy Hops grown. The first pocket arrived this week and realized £15 per cwt. The quality was exceptionally good for the first pocket. Picking will generally commence in about three weeks. Market quiet, with holders desirous of clearing out.—*Private Letter.*

### RECEIPTS, EXPORTS AND IMPORTS FOR THE PAST THIRTEEN YEARS.

| | RECEIPTS. | EXPORTS. | IMPORTS. | | RECEIPTS. | EXPORTS. | IMPORTS. |
|---|---|---|---|---|---|---|---|
| Year. | Bales. | Bales. | Bales. | Year. | Bales. | Bales. | Bales. |
| 1869. | 166,920. | 69,463... | 418 | 1876. | 84,138. | 46,116... | — |
| 1870. | 102,027. | 56,453... | — | 1877. | 84,358. | 44,493... | — |
| 1871. | 67,799. | 24,577... | — | 1878. | 138,160. | 78,949... | — |
| 1872. | 29,121. | 6,095... | 5,800 | 1879. | 93,480. | 34,749... | — |
| 1873. | 23,781. | 9,315... | 20,885 | 1880. | 82,608. | 43,954... | 2,772 |
| 1874. | 24,550. | 1,638... | 13,444 | 1881. | 96,988. | 43,027... | 2,094 |
| 1875. | 44,086. | 15,995... | | | | | |

### CROP PROSPECTS.

COOPERSTOWN, N. Y.—The weather for the growth of the hop in this county has been all that could be desired. The yards are generally looking well, and the quality bids fair to be of the best grade. Picking will become general next week, while several growers have already begun.—*Republican,* Aug. 24.

WATERVILLE, N. Y.—Dealers have nothing to do but guess and speculate on the incoming crop, which is looking better and better; and we have no reason to change our former estimate of 25 per cent. off of last year's crop in the state, and 10 to 15 in this vicinity, unless to make it a little more favorable. The outlook is a bright one for an average crop of best quality.—*Times,* Aug. 25.

COOPERSTOWN, N. Y.—There has been good progress made by the growing crop during the past week, and there is now a good prospect of a fine crop of hops. Allowing for the increased acreage, the estimated yield will be from 15 to 25 per cent. short of that of last year in the great Hop District of this state. On the Pacific Slope the yield will be fully up to the average. In Wisconsin and other western states the yield will be considerably under the average of the past two years. It is too early to judge of the English crop from reports at hand, but it is safe to say they will take several thousand bales of good American hops at what they deem a fair price. On the continent the yield is likely to fall considerably short of last year's crop. The price in this country will largely depend upon the extent of the English demand, for we shall have hops to export.—*Journal,* Aug. 20,

UTICA, N. Y.—The stock of last year's hops having been pretty well cleaned out. Utica buyers are now turning their attention to the crop now maturing. A few bales of the early hops have been sold to brewers from the Utica market during the past week at 36c. When asked concerning the coming crop this morning, a Utica dealer said it would have no superior as to quality among the various crops since 1869. The chances are that the 1881 crop will approximate in quantity that of 1880. In some of the principal districts of Oneida County the vine is heavily fruited, and the bearing is satisfactory in every respect. In the Stockbridge Valley the crop is not so satisfactory, but when compared with last year it is much better than had been anticipated. Reports from some points in Otsego County say the crop will almost, if not fully, equal that of last year. It is learned here on good authority that growers in Franklin County have contracted for the sale of their crops at 15c, which is beyond the estimated price of this section, owing to the abundance of hops.—*Observer,* Aug. 18.

### CASH PRICE CURRENT FOR HOPS IN NEW YORK.

| | | | CENTS PER LB. | |
|---|---|---|---|---|
| NEW YORKS, crop 1880, choice, | ... ... ... ... ... ... | 20 | to | 21 |
| "    "    medium, | ... ... ... ... ... | 18 | to | 19 |
| "    "    low to fair, | ... ... ... ... | 12 | to | 15 |
| EASTERN,    " | ... ... ... ... ... | 14 | to | 20 |
| WISCONSIN,    " | ... ... ... ... ... | 14 | to | 20 |
| YEARLINGS, crop 1879, | .. ... ... ... ... ... ... | 12 | to | 16 |
| OLDS, all Growths, | ... ... ... ... ... | 4 | to | 10 |
| PACIFIC COAST, new, | ... ... ... ... ... ... | 19 | to | 21 |
| BAVARIANS, | ... ... ... ... ... ... ... | 30 | to | 35 |

soon gained the edge and held favor through the 1870s and 1880s. In the Pacific Northwest, Corbitt and Macleay of Portland was probably the first commission house to enter the fields of the Willamette and Puyallup valleys.[19] In Seattle, Phillip Meis and Company was among the most prominent by the end of the century.

Commission houses were very much involved in collecting information about field conditions and in knowing the foreign as well as the domestic market prices. This led to the establishment of hop journals. Foremost was *Emmet Wells' Weekly Hop Circular*, published in New York, which provided the prices of hops from various locations, including New York, Wisconsin, Michigan, Bavaria, and Belgium, and later, California, Washington, and Oregon. Other articles dealt with shipments from foreign ports and accounts of European breweries. In addition to a number of domestic circulars, English commission houses also published helpful periodicals. For example, Thomas and Short of London distributed their circular along the West Coast in the hope of currying favor among growers there.[20]

The result of all this activity was that, well before the harvest, dozens of contracts were signed between the growers and the dealers, brewers' agents, and factors. In 1898, for example,

*3.4  Emmet Wells' Weekly Hop Circular, August 26, 1881. Begun in 1862, this serial was probably the most widely read of any among hop growers.*

*3.5  Hop sampling kit, made in New York City, undated. These hand tools were critical for determining the quality of the baled product, which determined the price. (Courtesy of John Alden Haight.)*

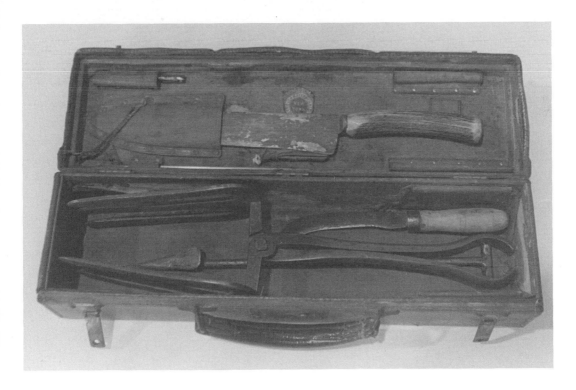

four growers in Marion County, Oregon, filed their contracts with the county clerk by July 1, and sixty-three were on record by August 15, with prices ranging from seven to ten cents per pound. In the early twentieth century at least a third of the Oregon crop was contracted before it was picked.[21]

The hop dealer was a shrewd businessman who was not likely to yield his advantage of an on-the-spot sale. Any number of ploys would be used to gain an advantage over a grower. The dealer generally had a central point in the hop district from which his agents would be sent. Sometimes the dealer sent a telegram to all of his agents commanding them to stop buying and come home to report. On receiving this dispatch the agents pretended to go homeward, but in reality they would take a circuitous route, examining all the hop yards in a predetermined area. An agent would present himself to the grower and say, "I have been out buying hops and would like to put one more lot in my report." Showing his dispatch to the farmer, he would add, "However, hops have fallen; so said the telegraph operator when he handed me this telegram. I have been paying forty cents; if you like, I'll take yours and report." The grower, believing that the agent was doing him a favor and that the price was a good one, took the 10 percent down that was offered and agreed to deliver the balance, when he would be paid in full. Only later, upon talking with his neighbors, did the grower learn that the same "favor" had been liberally conferred, and a few of them had held out for a cent or two more!

Agents might also gang up on a grower. The first agent would stop by and tell the farmer that he could not buy because hop prices had fallen. To be generous, however, this agent would offer the going market price. He then left. The next day, the second agent would stop by and offer five cents below the market price, seemingly sympathetic to the plight of the farmer in view of the uncertainties of the future, but he too left without consummating a sale. The grower, having the increasing impression that his crop would not fetch as high a price as he once thought, believed he could not refuse another good offer. When the third agent arrived, he came directly to the point and asked if the farmer had decided to sell his hops and if so, at what price. The farmer, of course, quoted the price he had been given by the previous two agents, and the third agent took the crop, paid him 10 percent down, and said he would let

TINGED WITH GOLD

him know when to deliver. Again, only later would the grower learn of the conspiracy.[22]

As everyone in the business knew, after the hops had been harvested, time was against the grower, for the hops lost value with each passing day. From the dealer's viewpoint, the best time to buy hops was at the end of the year, between Christmas and New Year's Day, when business was suspended. Farmers tended to be mellow traders during the holiday season, flocking to town to shop and stopping in at the office of the local hop buyer. Bargains were struck quickly when the farmer needed extra cash.[23]

However, for all the advantages of dealing with middlemen, hop growers often became disenchanted, especially if the offers were low. Dealing with an agent in upstate New York during the late 1860s, for example, generally cost the hop grower five cents more than if he were to take his crop to the New York City markets. Two cents went to the country dealer, transportation cost one or two cents, and the balance was lost in commission. As will be discussed further below, efforts to unite the growers and establish a constant price structure repeatedly failed in the East and were only partially successful in the Far West in the early twentieth century. Dissatisfied growers had little recourse but to initiate correspondence with some of the major brewers, such as John Taylor's Son; Coolidge, Pratt and Company; Amsdell Brothers; and John McKnight's Son, all of New York City. This might provide the grower with better access to the market, but then he also faced the problem of transporting his hops independently.[24]

By the late 1870s, after a few years of depressed prices and declining profits, hop growers began to raise a number of questions about the economic wisdom of sending hops to commission houses. "Would not all of our good hops have brought us ten cents, or more, per pound this year if they had been kept in our hop houses? In that case, all the dealers would have been compelled to send their agents into the country to secure them."[25] Once commission agents entered the industry, they began to sell consigned hops to the brewer for much more than he paid the grower. The agents might make as much as a dollar a bale, which the farmers believed was rightfully theirs. Equally apparent, often the commission agents did not have the money to pay for the hops but were merely providing a service. The farmers believed they were taking all the risks: "Many growers have got to sell their farms on account of low prices—*they* have done the work and the commission men have got fat off the farmer's industry."[26]

In reality, the grower was not relieved of market fluctuations; rather, middlemen had simply become part of the problem. Dealers regularly used their knowledge about the farmers, the soil, and weather conditions, and any slightly advance knowledge of market conditions, to their own advantage. Often they purposely overestimated the crop in a given area to suggest that there would be a large harvest and potential oversupply. The farmer, fearful of being caught holding hops, often agreed to early sales at a reduced price.

A large grower, like Patrick Cunningham of Mendocino County, might decide that a five-year contract with a San Francisco firm at the stable price of fifteen cents per pound was, on the whole, better than attempting to foretell the changes in the marketplace.[27] Under the terms of such a contract, a grower promised to deliver choice hops, and the merchant sold these in advance to other parties. Even if the grower did produce a splendid crop, however, the buyer protected himself with a contract clause requiring only the finest quality, and he reserved the right to reject hops for any of a number of reasons. Because sometimes a considerable difference of opinion arose as to the worth of the harvested hops, the contracts between the dealers and the growers led to litigation and court disputes.[28] In Salem, Oregon, for example, a grower named F. W. Buells refused to honor his agreement with a dealer, A. F. Blackhaus. In this case the contract provided that the hops were to be raised and sold by the defendant, that they were to be of "choice quality, sound condition, bright uniform color, fully matured, free from mould and damage from vermin, cleanly picked, well dried and cured, and put in good merchantable order."[29] The hops were to be delivered at Silverton during the month of October, but the dealer reserved the right to decide that if the hops were of inferior quality he would take the hops but make an adjustment in the price equal to the difference in their value. When a dispute arose about the value of the crop, the matter was taken to court. The decision of the judge was that the contract was a mortgage, and that the grower could discharge it by paying the money he was advanced with interest to the dealer. Both sides claimed victory,

# THE WORLD

PORTLAND, OREGON.

## OREGON HOP GROWERS' ASSOCIATION.

President—W. H. Egan, Brooks.
Secretary—James Winstanley, Salem.
Treasurer — Francis Feller, Butteville.
Directors—W. Goodrich, Chemawa.

### Hop Contract Decisions.

January 6, 1904, at Salem, before Burnett, J., a suit arising out of a hop contract was decided in favor of the grower. It was Geo. A. LaVie v. Walter L. Tooze et al.

The grower had contracted the hops to LaVie at 10½ cents a pound. He refused to deliver to LaVie, but delivered to Tooze. LaVie brought a replevin suit and secured possession of the hops. This the Supreme Court held could not be done upon a contract such as that under which LaVie claimed title, and the case was tried again to ascertain the amount which Tooze should recover because of the taking of the hops.

The jury brought in a verdict for Tooze for $1140.90, which is the value at 15¾ cents.

The average charge for recording hop contracts is about $3.75.

### HOP CONTRACTS.

L. A. Byrd, Jr., and T. A. Ditmars, from Fairfield, to T. Rosenwald & Co., New York, 8000 pounds nops at 17c. Signed January 12, 1904.

J. R. and A. L. Vanderbeck of Gervais, to T. Rosenwald & Co., 8000 pounds of hops at 17c. Signed January 12, 1904.

Joseph Reubens of Gervais, to T. Rosenwald & Co., 10,000 pounds hops at 17c. Signed January 12.

Theodore Reubens of Gervais, to T. Rosenwald & Co., 6000 pounds hops at 17c. Signed January 16.

H. H. Spaulding, Salem - Prairie, to A. Magnus Sons & Co., Chicago, 10,000 pounds hops at 16c. Signed February 22.

E. L. Pooler, Pratum, to A. Magnus Sons & Co., Chicago, 8000 pounds hops at 16c. Signed February 12 for three years.

J. A. Pooler and C. H. Lang, Salem, to A. Magnus Sons & Co., 22,000 pounds hops at 16c. Signed February 13, for three years.

J. R. and E. Coleman, and J. E. Forrest, Salem, to A. Magnus Sons & Co., 20,000 pounds hops at 16 cents.

Lillienthal Bros., to buy 10,000 pounds from A. C. Manning and George B. McClellan, Gervais, at 17½c.

Lillienthal Bros., to buy 10,000 pounds of Mike Keppinger, Gervais, 1904 crop, for 16c, and 10,000 pounds of 1905 crop for 14c.

Mrs. J. McKay, Moses McKay, I. F. Buysene, of Champoeg, to E. Wattenberg & Co., New York, 8000 pounds of hops at 17c.

J. A. Pooler and C. H. Lange to A. Magnus Sons Co., 22,000 pounds for 3 years at 1*c.

E. L. Pooler to A. Magnus Sons Co., 8000 pounds at 16c for 3 years.

H. H. Spalding to A. Magnus Sons Co., 10,000 pounds for 3 years at 16c.

J. R. Coleman and others to A. Magnus Sons Co., 20,000 pounds at 16c.

Theodore Rubens to T. Rosenwald & Co., 10,000 pounds at 17c.

J. R. Vanderbeck to T. Rosenwald & Co., 8000 pounds at 17c.

L. A. Byrd, Jr., to T. A. Ditmars, 8000 pounds at 17c.

Wm. J. Gulden, Salem, to George A. LaVie, 8000 pounds at 16c for 1904, and 12c for 1905-6-7-8.

Quong Hing, Salem, 40,000 pounds at 16c, to Krebs Bros., Salem.

J. H. Burton, Independence, to Horst, Lachmund Co, of Portland, 20,000 lbs. at 10c. Signed December 21, 1901.

W. D. Huston, of Suver, to T. A. Livesley & Co., of Salem, Oregon, 10,000 pounds at 12½c. Signed January 23, 1903.

J. T. James to Benj. Schwartz &

*3.6* The World *(Portland, Oregon), c. 1905, p. 1. Published by the Oregon Hop Growers' Association, this issue focuses on hop contracts and hop contract decisions in the courts. (Oregon Historical Society, Portland, Oreg.)*

but the grower stood to lose more. As one editorial in a California agricultural paper put it, the tendency of hop growers to contract portions of their crops to dealers in San Francisco was "simply suicidal."[30]

In Washington, an act of the legislature in 1899 gave the governor the authority to appoint a hop inspector. This was not a new idea, but it is curious that growers and dealers viewed it as a mechanism that denied them their *privilege* of going into court when a dispute would arise.[31]

Whereas a single, discontented grower had little recourse in the face of much larger economic issues, an association of growers might be able to solve some of them. Early efforts on the part of growers to organize for the promotion of their common interests met with little success. In New England, no formal effort has come to light. The limited size of most fields and the small number of growers may have worked against their forming a common front. By the late nineteenth century, however, the widespread success of the Patrons of Husbandry and other fraternal organizations spurred hop growers west of the Hudson River to establish an association that would advance their common interests.

In New York State, growers began to organize by the middle of 1874. Calls for a union that would serve primarily as a forum for discussion became commonplace. The first general meeting of the Hop Growers' Convention in the Waterville area was set for Saturday, October 16, 1875, with the purpose of adopting a constitution and setting an agenda. Although the growers were off to a good start, their activities amounted to little more than campaigning for a standard-size hop box and compiling annual production statistics.[32] Despite the obvious advantages of acting in concert to fix the price, growers in the state failed to act in unison and thus continued to be completely at the mercy of the market.[33] For years, upstate growers met in conventions at Norwich, and later they gathered in annual social outings at Sylvan Beach. However, the thousands who assembled there were seeking a relaxed atmosphere, and there was little, if any, discussion of economics.[34]

In Wisconsin the effort to establish various associations had only begun to get under way at the peak of production in 1868, when prices bottomed out. Growers in Sauk County and Juneau County, Wisconsin, began to organize, but interest in

TINGED WITH GOLD

# Hop Growers' Convention !

PURSUANT to the action taken at the Meeting of hop growers held at Oneida in August, a Convention of Hop Growers will be held

## At Granger Hall, Waterville,

### Saturday, Oct. 16, 1875,

at 10 o'clock A. M.  The business of the Convention will be a report from the committee on organization, the adoption of a Constitution and By-Laws, and discussions on various topics.

By order,

78w1                          M. W. COLE, Secretary.

3.7  The first public notice of a growers' movement to organize. (From Waterville Times, October 14, 1875.)

growing the crop died so completely that nothing was accomplished.[35]

In California, by contrast, the early attempts at organizing growers were more impressive. In part, this was related to the rise of fruit-growers' associations. Looking at parallel efforts in related agricultural industries, the Mendocino Hop Growers Association actually attempted to control the price that would be paid for hop picking and the prices that they would receive for their product. In 1891, because of concern that large fruit and grain crops would probably make pickers scarce, growers fixed prices at $1.10 per hundred pounds for those who came into the field at their own expense and paid $1.00 per hundred to those who were transported to the field at the expense of the grower.[36] The Sonoma County Hop-Growers Association also fixed a uniform price, although not every grower joined the organization. Appealing to those unaffiliated growers in the region who were gaining all the benefits of the association without any effort, President Guy E. Gosse remarked, "Gentlemen, in union there is strength, and another invitation is hereby extended to you."[37]

In Oregon growers first attempted to organize in 1877, but without success. The extended efforts that occurred during the mid-1890s were somewhat localized. The Hop Growers Association of Lane County proposed to assist in picking, curing, and storing the hops of stockholders, besides doing a general warehouse business. This promotional organization involved about forty growers who decided to seek incorporation in August 1893.[38] At this point an upward swing in prices, the effect of advertising that there was "money in hops," and reports of a "splendid outlook" led to an increase in the number of acres under production. Some members of the association who had decided to pay thirty-five cents per box for picking found that others had decided to disregard the agreement and pay more, hoping to attract the best pickers as quickly as possible. It was apparent that in a bull market the association was powerless.[39]

Elsewhere in Oregon a group of hop growers in the Independence vicinity decided to organize a permanent association in August 1894. Its purpose was "to advance the best interests by meeting from time to time to discuss the best method of harvesting and cultivating." Right from the start, however, members of the association agreed that the group would not have the power to dictate to any member where, how, or to whom he could purchase his supplies or dispose of his crop. The relatively high prices offered in the early 1890s made these growers shy away from any attempt to limit the realization of their maximum potential. Its sights lowered, the Willamette Hop Grower's Association, as it came to be known, met only infrequently. Its chief accomplishment may have been setting a general price per box in a given season.[40]

On the other hand, the various organizations did serve to voice the general alarm when the price of hops dropped in the late 1890s. In October 1899 the Oregon Hop Growers' Association was formed at Woodburn, with the idea that all the hops of the smaller growers would be bought up and held until the prices increased. Other purposes of the association included building, acquiring, or leasing warehouses, lending money and extending credit to members, borrowing money when necessary, and attempting to influence railroad shipment rates.[41] This was a concerted effort to meet the mounting "evils" being perpetrated by the commission agents. At the outset the organization controlled almost half of the output in the state. Nearly all the growers joined in the hope that the association would control the price, and a plan was adopted for the entire crop of the members to be turned over to a single agent. The plan was never carried out, however, for several growers were unwilling to turn over their product without strings attached, and in 1900 the price for Oregon hops doubled. Five years later the association dissolved.[42]

In 1902 M. H. Durst of Wheatland, California, wrote a widely publicized open letter to Pacific Coast hop growers. Upon returning from a trip to England in April, he was sorry to learn that many growers had already signed contracts with dealers for ten to twenty-one cents. Because the weather in Europe was not favorable, and because brewers there and in the United States were beginning to run low, a keen demand would drive up the prices. He advised growers to cooperate, hold their hops beyond the harvest for two or three months, and refuse to sell at anything under twenty-five cents. Durst believed this was a "grand opportunity for a second co-operative effort by the hop growers of Oregon" and hoped to see the movement realized.[43]

Durst's actions were criticized as self-serving because his fields were some of the largest in the country. More often than not, large growers contracted directly with one or more brewers. In April 1904, for example, the entire product of the California ranch of E. Clemens Horst was sold to a brewery in Milwaukee, a contract of 10,200 bales of hops, or over twenty million pounds.[44]

Growers also became convinced that buyers were getting rich at their expense, and they attempted to devise a means by which middlemen might be eliminated. This was one of the explicit complaints that resulted in the formation of the third Hop Grower's Association, organized in 1914. In this case the group intended to buy at wholesale all the supplies necessary and distribute them to members at the lowest possible prices, and to pool their crops and have the association sell directly to brewers.[45] Unfortunately, World War I disrupted the market and England declared an embargo on U.S. hops, so the idea went largely untested.

Similar organizations were founded in Washington. A meeting of the Puyallup growers was held in the store of Ezra Meeker on August 11, 1877, to discuss the means by which they could assemble the estimated number of pickers that were needed. It was the position of the growers that, unless they could obtain better rates than were being offered by the Northern Pacific Railroad, they would look for pickers not in Oregon but in British Columbia. There, about twelve hundred workers recently engaged in the salmon canneries were unemployed.[46]

The Puyallup growers were also conscious of the exorbitant rates charged for the transportation of their baled product. Before the coming of the railroad in the mid-1870s, it cost five dollars a ton to haul their hops to Tacoma by wagon, and another twenty or twenty-five to ship them to market in San Francisco. The railroad picked them up from stations in the valley and placed them in San Francisco for six dollars per ton.[47] Although, by comparison to the prices they had paid previously, the recent situation was a marked improvement, the effort of the Puyallup growers to establish an alternate line continued for a number of years.

The early history of the organizations on the eastern side of the Cascades is unclear. Yakima Valley growers organized as early as 1894, but there is little record of continued activity.[48] Their primary concerns were suggested by the two standing committees, the first for securing pickers and the second for the end-of-the-season barbecue. The Southwestern Washington Hop Growers' Association was founded in January 1900 in Chehalis.[49] The members controlled over half of the acreage in that part of the state. Its purposes were modeled after the statewide association in Oregon.

The call to unite all hop growers on the Pacific Coast was heard once again when, after a period of relative prosperity in the twenties, a severe downturn occurred in the mid-1930s. This was undoubtedly the greatest period of organizational activity among growers in the United States. The Yakima Valley Hop Growers Association was formed in 1932 with the objective of securing an agreement among growers about the wages that would be paid to the pickers. The Oregon Hop Growers Association was begun in 1932 as well, with headquarters in Salem, and the United Hop Growers of California was organized early in 1933. However, the attempt to establish the U.S. Hop Growers' Protective Association, with its headquarters in Seattle, proved unsuccessful, despite the fact that committees for cooperative control were formed in California, Oregon, and Washington.[50] In the face of widespread economic difficulty, most growers did not respond. With the Depression causing an increasing amount of desperation, the inclination to grab any kind of contract was overpowering. The small growers hoped to do more of their own work and drive out the big producers, while the large growers believed they had enough capital to keep going at the expense of those with small yards.

TINGED WITH GOLD

Only in New York was there a modest amount of cooperation. There, growers appealed to the legislature, and funding was provided for the building of cooperative kilns and for the appointment of a commission to study the diseases in that state.[51] Far fewer in number, New York growers formed a cooperative during the summer of 1937. Only a little over two hundred acres were planted, however, a far cry from the halcyon years of the late nineteenth century.[52]

In early 1938, thanks to the intervention of representatives from the Pacific Coast states, the United States Congress passed and the president signed legislation that included hops among the commodities to which marketing agreements applied. Growers, dealers, and brewers on the Pacific Coast were

### Dear Mr. Hop Grower:

Conditions in the hop business are going to be worse, unless we do something about it. . . . and that is the purpose of this letter.

For several years, the amount of hops harvested has exceeded the demand until at the present time, according to government figures there are some 117,000 bales in the hands of brewers, brokers and growers. Reliable estimates show that about 196,000 bales will be produced this year, making a total of about 313,000 bales available to brewers.

Yet unless the sale of beer takes a rapid jump, only 180,000 bales of hops will be used this coming year. Obviously, this is a bad situation. BUT it isn't hopeless. That is, if we act and act right now.

By leaving a portion of our hops on the vines, by all of us sharing alike, we can in time remedy the situation. And that is exactly the purpose of the Oregon, Washington, California HOP MARKETING AGREEMENT. Under this agreement all growers in the three states will make proportionate reductions in their crop until production meets the demand. By this method, everyone will be protected from serious price reductions.

The agreement will be administered by the growers themselves—someone from your community whom you've helped to elect will sit on the board and decide all matters of policy.

Soon, you will receive a ballot asking whether or not you wish this agreement. Vote "YES" and send it in immediately. If you vote in the negative, you will be voting for ruinous prices and bankruptcy for many growers.

The Time Is Short---SEND IN YOUR BALLOT AS SOON AS YOU RECEIVE IT!

## Vote (YES) for the Oregon-California-Washington HOP MARKETING AGREEMENT

3.8 Advertisement for the Oregon-California-Washington Hop Marketing Agreement, July 1938. Overproduction led to an attempt on the part of growers to make proportional reductions in their crops. (From Pacific Hop Grower 8, no. 3 [July 1938]: 3.)

forced to work together toward a hop-marketing agreement.[53] Growers in all three states made proportionate reductions in their crop until the supply met the demand. Large growers, including Lloyd L. Hughes, Incorporated, and the Yakima Chief Ranches, sued in federal court, which declared the law unconstitutional because a provision of the agreement, regulated by the Hop Control Board, required all growers to obtain certificates before they could dispose of their hops.[54] The order expired in 1945 but was reinstituted in 1949 as the Federal Hop Control Board. Its sole function was to stabilize the market by equalizing supply and demand. At last, or so it seemed, hop growers had a device to counter market fluctuations.[55]

If the grower had overcome all the problems of raising hops and consigned his crop to a dealer for a reasonable price, the thorniest problem that remained was that of collecting the crop. This required more hands than were usually needed on the farm or ranch and, as mentioned above, this was a major expense. Perhaps twenty pickers would be needed to harvest an average four-acre field, and they were often in high demand. As the fields expanded, more labor had to be drawn from outside the immediate area. Inevitably the grower had to attract pickers he did not know, as well as face the problems of feeding and housing them. The larger the number of pickers, the greater the difficulties became.

During the late eighteenth and early nineteenth centuries in New England and in New York the scale of operations was small enough that the need for additional labor was satisfied by appeals in adjacent towns and villages. Many of these pickers, who were employed for a few days, could walk home for lunch and sleep at home. Even with such local help, the grower often provided the pickers at least one meal a day.

Food preparation was no small task. Butchering and baking would take a considerable amount of time, for pounds of meat and loaves of bread had to be near at hand. The greatest portion of the work preparing for the pickers fell on the shoulders of the farmer's wife and daughters. Occasionally, an additional domestic servant was hired to spread the work load.

In the cases where the hop grower was providing housing and meals, the preparation for the pickers began well in advance. It would often include cleaning up the dormitory areas or sleeping rooms and assembling sufficient bedding and

TINGED WITH GOLD

fuel for the cooking stoves. Washing required extra tubs and towels.

Given a harvest season that was only three weeks long, the grower generally considered the crowded conditions only a temporary inconvenience. It was, nevertheless, uncomfortable. As many as two dozen people, in addition to the grower's family, would sleep in a two-story farmhouse.[56] As Mary E. Tooke, who watched thirty glass workers from Oswego move into her cousin's house near Bouckville, commented: "It's like Bedlam let loose the way they enter and take possession of the whole house."[57]

As production increased, growers began to depend upon pickers in nearby cities for the extra help necessary to harvest the crop. The larger growers, such as James F. Clark of Otsego County, were faced with the necessity of taking mea-

3.9  *James F. Clark and his family, Cooperstown, New York, undated. With hop bouquets in hand, the family was photographed in the fields at harvest. (New York State Historical Association, Cooperstown, N.Y.)*

sures to accommodate dozens of pickers.[58] Clark began to raise hops about 1876, and he erected a single building to house his city pickers. By 1881 he employed about 150 workers, and a few years later he erected, on the River Road leading from Cooperstown to Phoenix, an entire settlement that became known as "Hop City" to accommodate an expanded work force. During the harvest the working population in this temporary village was nearly a thousand, about three hundred of whom were local residents (primarily superintendents and field tenders) who boarded at home. Most of the local pickers came from Cooperstown, Phoenix, Hop Factory, Hyde Park, and Toddsville, and were transported to and fro by Clark's teams.[59] The remainder, primarily from Albany, arrived by rail, disembarking at the station; they too were carried by the wagonload to the fields.

The layout of Hop City is worth noting for its orderly arrangement. It was somewhat like a small, paternalistic New England mill village; Clark's residence, a comfortable dwelling, stood at the head of the main avenue. On both sides of the thoroughfare were the seven hop houses, connected by tele-

*3.10 "Hop City," near Phoenix, Otsego County, New York, c. 1893. The center of James F. Clark's fields as it appeared near the height of its development. (New York State Historical Society, Cooperstown, N.Y.)*

TINGED WITH GOLD

phones. Each hop house had two kilns, for a total drying capacity of from fifteen to sixteen thousand pounds in twenty-four hours. In the first hop house was located the office of the mayor of Hop City, chosen from among the pickers, and a barber shop. There the mayor transacted all business. Scattered along the rear of the kilns were four substantial frame structures for sleeping and eating, and a number of smaller dwellings to house the large force of domestics that fed the city pickers. Four wells, driven by windmills, provided fresh water. Hop City also included a blacksmith shop, a cobbler shop, and a store where gloves and tobacco as well as other necessities could be purchased.[60] Clark recognized that hop growing on a large scale required a significant investment to accommodate the number of seasonal employees.

Perhaps because of the widespread advertising that attracted legitimate pickers, a number of vagabonds began to take up residence in the Cooperstown region. In fact, during the 1870s perhaps no other community in the state was concerned with the "tramp nuisance" as much as Cooperstown. With a genteel tradition and a rising number of summer visitors, residents looked with increasing alarm at the number of vagrants who were camping almost everywhere around the village. The matter came to a head on Monday evening, August 26, 1878, when citizens filled the courthouse and aired their complaints to the village board of trustees. They declared an urgent need for a police force to protect the citizenry "against the aggressive arts of a class of tramps engaged in assaults on persons and property,"[61] and a lengthy discussion with many of the major hop growers ensued. As a result, a resolution was adopted by which the trustees were requested to employ a day and night watch of sheriff's deputies during the harvest season. In addition, all idle tramps were to be warned to leave the village, and if necessary they would be arrested and imprisoned for ten days, being fed only bread and water.[62]

Perhaps more effective than the deputies were the circulars that were soon posted about the area, requesting all citizens to notify the sheriff or police of the presence of tramps. Most effective, however, were still other notices. "Buy a Revolver and Shoot the Tramps" crowed an advertisement by a local grocer, who had a full line of sportsmen's goods on hand.[63]

For a few years, such scare tactics seemed to work. By 1881,

3.11 Evidence of the concern about vagabonds camping in the Cooperstown hop-growing area. (Freeman's Journal, September 19, 1878.)

however, the problem had returned. The citizens charged that growers were bringing into "the country many dissolute men and women, with whom our home pickers do not care to come into contact, either in the field or at the hop dances, and whose presence in these rural districts, even for a few weeks, is in every way undesirable." A "long list of petty crimes, and offenses against persons and property," during the harvest season was cited to prove the evil effects of these transients.[64]

Although vegetable gardens and orchards did suffer at harvest time, the "tramp nuisance" was almost always exaggerated. "The gang of seventy-five tramps quartered in a swamp and camping out like gypsies is a myth," wrote one observer from Utica in the late 1870s.[65] Hop picking was too much like work for the average tramp. The few tramps that were seen became obvious because they seemed "always on hand at meal times, taking the first seat at the table, and promising to work after dinner, or refusing to leave until they have had a meal."[66] However distasteful their presence might be, particularly to the women, there was probably little reason to fear these knights of the road.

While Cooperstown residents were annoyed at having too much "labor" in the vicinity, elsewhere in the country the need for extra hands was a much more serious business. In the West some farms were very remote, and growers regularly hired agents to recruit pickers. The search extended well beyond the towns in the vicinity to native American reservations, where bargains were struck with tribal leaders. Large growers on the Pacific Coast advertised as far east as the Mississippi Valley.

Occasionally the pickers were a little slow in assembling, and growers panicked and began to search for any labor pool close at hand. In 1893, for example, to help meet the need for labor and to avoid employing Chinese, the directors of the school system in Independence, Oregon, decided not to open the schools until the latter part of September, so that the children could help in the harvest.[67] Such decisions were much more commonplace in the agrarian economy of the period, when the growers' needs were associated with the prosperity of the community, than they are today.

As in the East, growers in the West used the railroad to bring in a supply of labor wherever possible. In 1904, for example, a North Yakima firm of hop growers sent to the employment bureau in Tacoma a request for 150 pickers. The firm

TINGED WITH GOLD

offered to advance the fare of the entire party from the sound to the field and to charter a special car for the trip.[68]

The need for pickers remained acute among the large commercial operators on the Pacific Coast well into the twentieth century. In 1935 growers turned to the local U.S. Re-Employment Service and the Marion County (Oregon) Relief Administration to recruit help. The governor even pleaded over airwaves for help in finding an estimated six thousand additional pickers. Suggestions from growers in Oregon to use members of the Civilian Conservation Corps were rebuffed, for administrators doubted the situation was a "real emergency."[69] During World War II prisoners of war were put to work in the fields.

The use of outside labor inevitably led to the need for increased policing. There were two kinds of infractions: the first concerned the quality of the work in the fields, and the second arose from the social problems that, generally speaking, only a few pickers would create.

Perhaps the most frequent difficulties derived from "dirty picking." Growers and their field staff continually needed to remind pickers that they were being paid to pick hops, not vines, stems, and leaves. Occasionally a confrontation occurred due to the differences of opinion about the standards being applied.

Social policing was much more difficult for the grower. Young pickers, some of whom were away from home for the first time, would test their strength and the tolerance of the public with antics that were annoying and, at times, dangerous. Despite posted warnings, for example, frolicking pickers repeatedly tore off the tops of the vines and drove through the streets with their horses and carriages covered and themselves decorated with the crop. Accusations of petty larceny were commonplace in the camps. Hop pickers were sometimes charged with stealing fowl or livestock, or even the wash tubs, washboards, washing machine, and clothesline of a neighbor. Further, the grower often had to deal with accidents and sometimes with a death by drowning in a river bottom or nearby pond.[70]

Many of the problems resulted from an overindulgence in alcoholic beverages. As the number of city pickers increased, the townsfolk began to perceive the growers as harboring people of poor character.[71] Although most of the city pickers were temperate and law-abiding, a few were not. Not every grower was so conscientious as to forbid "men and women of degraded character" from entering his yard. Hence, the question arose, "Would any right-minded parent want a daughter of his to go out from the influences of home among such influences as these?"[72]

Far more difficult than dealing with the general discontent of the townsfolk was the problem of controlling dissatisfied pickers. "Despicable villains" would enter hop yards and cut a number of vines from the poles, destroying them before the end of the harvest.[73] More serious than the damage done to the vines were the fires that destroyed hop buildings and at times threatened to burn entire villages. For instance, when C. L. Terry of Waterville lost two hop kilns, a barn, and the boarding house for his pickers on Sunday, September 8th, 1874, no one seemed very curious about the cause of the fire. Terry may have taken comfort in the fact that the structures were insured, and he could carry on using his father's facilities next door. But fifteen days later, about midnight, the elder Terry's hop house also caught fire, consuming an entire crop of late hops. This time the fire was more closely investigated as a possible case of arson. Unfortunately, neither the structure nor its contents was insured.[74]

Growers learned that the field hand who was chided or discharged by an overseer would bear watching, for a single match dropped into the hop bin was an easy form of revenge. At any time during the drying process or thereafter, that match could ignite and within seconds cause an uncontrollable fire.[75] One of the most memorable events on the Durst Ranch, in Wheatland, California, was a fire in September 1889, which destroyed a triple kiln and two tons of hops. It was believed to have been started by matches enclosed in the bales by discontented workers. As late as the 1930s, growers who attempted to make their field hands pick more cleanly faced both labor riots and strikes.[76]

For many commercial growers, dealing with the transient labor became a necessary evil in the business. Inevitably there were differences of opinion between worker and employer, and the specter of mob violence—relatively unknown in the early eastern fields, where there were few farms requiring a large number of pickers—became a recurring nightmare.

Commercial growers also had a dream: a machine that would relieve them of the necessity of dealing with pickers. This not

TINGED WITH GOLD

only would mitigate the personnel problems but also would reduce the considerable cash outlay that was necessary every year. Growers became excited when mechanics brought forward their inventions, because they regarded a mechanized processor as an investment and as an appropriate avenue for research and development.

Many of the early inventions were limited in scope and merely dealt with improvements in manual picking. For example, William Brooks, an Otsego County native, invented a "hop screen," which was placed over the hop box and supposedly allowed only the hop to pass through, leaving the leaves and stems behind.[77] However, the mechanical picker, a machine that would pick only the hops from the vine, was the true goal. In 1868 a trial of mechanical hop pickers in the fields south of Newport, Wisconsin, brought out about two hundred spectators. The contest pitted the "Watertown Machine" against the "Paddock Machine," built in nearby Baraboo. For both machines the vine had to be cut down, chopped into pieces, and presented to the mechanical picker. Both were composed of spiked rollers for picking and a cylindrical screen for separating the burrs from the leaves. And both inventions left some hops on the vine, often mangled those that were picked, and required such care that they could be used only very slowly. The result was declared a draw. The inventors received little encouragement from the growers, who doubted that either machine would be of use. In fact, within a few weeks it was reported that all the new devices had been set aside in the rush to pick the crop.[78]

The dream of a machine that would pick hops cleanly and efficiently continued to surface. Ten years later, H. Niles Harrington and Charles Osborne of Peterboro, New York, invented a hop picker and separator. A stationary device, it required the vines to be cut from their poles, sliced into three- or four-foot segments, and fed into the picker, while the hops, discharged into the separator under the picking table, were fed into sacks. The machine was propelled by one man; two to four others fed it.[79] It too was field tested and abandoned as unsatisfactory. Yet another processor was "perfected" by George Spencer of Eaton, H. G. Locke of Waterville, and George Lawrence of Madison, New York. It was "about the size of an ordinary clothes wringer, [was] propelled by means of a treadle, and [ran] as easily as a light sewing machine."[80] Although

Locke traveled widely, even visiting Europe to promote the use of this invention, it does not seem to have been adopted in the field.

The first successful inventor was a western grower who had the means to persevere and, because of the size of his fields, a serious need for a dependable alternative to hiring a large labor force. In 1909 E. Clemens Horst, the well-known California grower, invented a stationary machine that he claimed could replace 450 hand pickers, even though it required the entire vine to be cut, loaded in wagons, and brought to a single location. The machine was about fifty feet long, fifteen feet wide, and ten feet high. The vines fed into one end traveled over two sets of revolving drums equipped with V-shaped fingers that plucked the hops free of vines, stems, and leaves. A number of conveyor belts moved the freshly picked hops. It was claimed that two men with one machine could process one hundred pounds per minute![81]

The use of stationary pickers was largely confined to commercial operations and required a significant investment, but on the whole it could be considered a success. A less expensive stationary picker-dryer was invented about 1928 by R. Langevin of White Bluffs, near Yakima. In this instance, thirty men with the aid of a fifteen-hundred-dollar machine were supposed to be able to do the work of two hundred men.[82] The vines were cut in the fields and carried to conveyors at the mouth of the two-story picking machine. Hop vines with the cones entered one end of the picker, which was a maze of gears, chains, and belts. Steel fingers on rotating drums stripped the cones from the vines, which were diverted as waste and hauled back to the fields as humus. Stems, twigs, and leaves were removed by a combination of vibrating screens, air drafts, and human pickers, before the hops were sent off to the kilns.

The idea of a machine capable of picking in the fields, further simplifying operations, was not realized until Edward Thys, the son-in-law of E. Clemens Horst, invented a portable picker in 1937. The Thys picker was towed through the fields by a tractor and addressed two rows at a time, leaving the vines in the field. It had a capacity of about two acres per day, or between five and eight hundred pounds of hops per hour.[83] Using such a portable picker, it cost less than two cents per pound to harvest the crop, although stationary mechanical pickers remained the most economical.

TINGED WITH GOLD

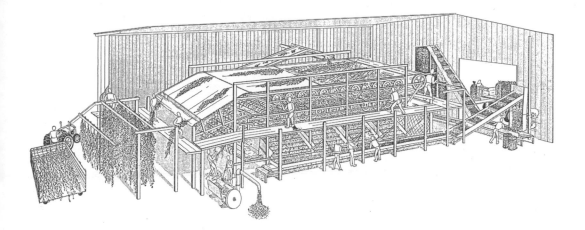

The use of mechanical picking machines became the topic of considerable discussion among Pacific Coast growers, largely because of a promotional film that showed them in operation.[84] Not only would hop-picking machines simplify growers' lives, but during World War II, when growers suffered from a drastic shortage of field labor, mechanical picking seemed to be the only alternative. After the war it soon became clear that the days of the seasonal laborer in the hop fields were numbered. By 1950 it was estimated that about 85 percent of the hops in the United States were picked by machine. Since then, the shift from the worker in the field who cultivated and harvested the crop to the mechanic who tends a machine has moved the modern hop farm into the realm of a hop factory.[85]

The concerns of the grower—market fluctuations and recurring labor shortages—were repeatedly addressed, but only in the twentieth century were solutions found. Market fluctuation was complicated by middlemen and led to feeble attempts to control prices. Growers' efforts at self-control were generally unsuccessful. The recognition of this fact, in turn, led to government regulation and reinforced the trend to an ever more specialized industry. Labor problems encouraged the introduction of mechanical pickers, a significant investment even for large growers, permanently altering the nature of the harvesting process and diminishing the need for the auxiliary structures associated with human pickers.

*3.12 The inner workings of the typical stationary picker, 1954. (From Anglo-American Council on Productivity, The Hop Industry [London and New York: Productivity Team Report], figure 13.)*

# 4

## The Pickers

Just as the lure of money was responsible for attracting growers of hops, the idea that a sizeable amount of cash could be made with a modest amount of effort led many people to try their hands at picking. In addition to the monetary benefits, advertisements encouraged prospective new pickers by emphasizing that picking hops was healthful. The sunshine and pure country air were invigorating, the outdoor exercise was light, and the season was so short that no one could labor too long in the fields. "Those who are habitually engaged in hop growing have been so uniformly in good health, as to attract the attention of medical philosophers," claimed one early New England agricultural journal.[1] The message provided a ready justification for joining the harvest simply to enjoy a change of scene.

From the picker's point of view, after the decision had been made to join the harvest, the principal questions were how to get to the best fields, and where to find the best working and living conditions. Life in the yards and in the camp was important because the ratio between local folk and "city pickers" varied considerably, as did the "class" of persons and their ethnic composition.

In the late eighteenth and early nineteenth centuries, commercial hop farmers in New England and New York expanded their labor force at harvest time by recruiting the residents of nearby farms and villages.[2] The hop grower preferred these "home pickers," because often they were neighbors who could be held more accountable for their actions. On their part, home pickers often felt more comfortable in their surroundings and often returned to the same field every year, furthering their bond with a particular grower. In some instances the families that worked in the fields were related, and they often joined together to help one another.

Early growers in the Northeast also preferred hiring the youth of the community, primarily because adult pickers were more expensive.[3] Hop picking was not a task that required tremendous muscle. Dexterity and perseverance were the keys. As each fragile cone represented a fraction of a cent, it offered profitable work for those with swift fingers.[4]

In the East, women performed much of the field labor in harvesting because most men could not leave their farms or factory jobs for the entire harvest season. In general, women and girls were believed to be better pickers, and they were

4.1 Hop pickers, pole
puller, and field boss,
Cazenovia, New York,
c. 1880(?). (New York
State Historical Associa-
tion, Cooperstown, N.Y.)

willing to work for less.[5] The mood among the pickers was often
one of a half-holiday, and the season involved a considerable
amount of clothing preparation and social intercourse. As one
Otsego hop grower wrote at mid-century, "I have often heard
the remark, that our hops are all picked by females, many of
whom spend weeks in preparatory making dresses, sacks, bonnets, gloves, & c., thinking nothing of the time thus spent, if
they can only have a good time picking."[6] A farmer generally
sent around a wagon early in the morning to collect home
pickers, chiefly women, who would climb aboard in their old
clothes, with their bundle of gloves, stocking bags, sun bonnet
or broad straw hat, bottle of tea, dinner basket, and perhaps a
three-legged stool. Young children would be brought along, but
they soon tired of any repetitive chores and were allowed to
amuse themselves. Alternatively, the pickers might organize
themselves into groups, hitching up a wagon or cart in the
village, loading aboard the day's provisions, and setting off for
the fields before daybreak.

Although a considerable number of home pickers continued
to work during the hop harvest, a change in the labor force
became noticeable about mid-century. As the hop fields of

central New York expanded, growers found it necessary to increase the number of pickers in order to complete the harvest in a timely fashion. More hands from outside the immediate area were hired. By 1863 growers or their agents searched for pickers not only in the towns and villages of Oneida, Otsego, Madison, Chenango, and Herkimer counties,[7] but also in nearby cities such as Syracuse and Utica, and towns along the railroad as far as Albany.[8] The same activities could be seen in the hop-growing region of Wisconsin. Farmers first depended upon the pickers from Loganville, Reedsburg, Sauk City, and Watertown, but soon extended their search to Monroe, Janesville, and Milwaukee.[9]

The role of the agents in enlisting these small armies cannot be overlooked. In 1878 agents in upstate New York received fifty cents per head, going house to house in a number of cities and towns long before the harvest to engage help.[10] Growers preferred families or young people, for the image of the home picker did not fade easily. Proper young ladies would be assured of supervision and could look forward to the outing as a chance to meet an eligible bachelor. However, young men who performed the heavier tasks of pole pulling and hauling hop boxes were in short supply, particularly during the Civil War.[11] In some cases someone who had worked for a particular farmer during the previous year would be in correspondence with him and could serve as an organizer, informing prospective pickers of the transportation arrangements, the nature of the accommodations, the food, and naturally, an idea about the rate of pay. In these instances the farmer might send his field manager with a wagon to the local train station to meet the party of pickers, often the Sunday before the harvest began.

In fact, local railroads took advantage of the opportunity to work with the farmers and dealers and ran special trains to and from the stations nearest the hop yards. For example, the first passenger train on the main line of the New York and Oswego, later known as the New York, Ontario and Western Railroad, was conducted between Oneida (Madison County) and West Monroe (Oswego County) on August 29, 1869, for the express purpose of bringing the hop pickers.[12] In later years railroads extended the routes to gain access to larger labor markets. At about the same time, in the hop-growing district of Wisconsin, the Milwaukee and St. Paul Railroad brought pickers from the east at full fare but gave free tickets to return.

4.2 *Hop pickers, Coopers-town, New York, vicinity, before 1900. A forest of poles laden with hops is gradually being harvested while a collection wagon makes its rounds. Taut cloths provide shade over the picked hops to prevent wilting. (New York State Historical Association, Cooperstown, N.Y.)*

Pickers often found it harder to return home than to arrive, for the railroads were not as anxious to be accommodating. Often, pickers found they had to stay over an extra day or two.

In Wisconsin the railroads transported approximately eight thousand pickers, roughly a third of the estimated total labor force.[13] This was about half of the pickers who came from outside the hop-growing region.[14] Elsewhere the need for labor was even greater. In the late 1870s "city pickers" made up about three-quarters of all the field hands involved in the central New York harvest.[15] Trains bearing city dwellers arrived at rural stations during harvest season, often causing considerable excitement. Hundreds of strange people had to find their trunks and baggage, and the right wagon destined for the right grower. Some pickers returned to the same fields every year, but many more merely followed the advertisements without a specific commitment. A more general approach might suffice: simply walk in whatever direction promised the greatest reward for one's effort. Minnie Henn, of Central Square, Otsego County, was typical of the picker who annually returned to Madison County. "When I was 13 years old, 1892, my brother Oscar Henn, and I went hop picking to the Porters'. . . . The next year, 1893, we went to Charles Bush. . . . The next year, 1894, when I was 15 years old, I went to the Bridge farm," and "the next year, 1895, I went to Stringers'."[16]

In some instances pickers seem to have found the agents rather than the reverse. Agents like M. Loewenmeyer, representing C. G. Mueller of Milwaukee, would station themselves during the summer at the hotel or at one of the hardware stores in a marketing town, such as Kilbourn City, and advertise their readiness to provide the local growers with an adequate number of hop pickers. In nearby Baraboo, LeRoy Gates reputedly engaged all the women within a hundred-mile radius. When, at the end of August, the agents signaled the need for additional help, wagon and trainloads of pickers arrived. Kilbourn City had a population of only fifteen hundred inhabitants; within a week thirty thousand pickers arrived, transforming the small town into a teeming municipality. As the local newspaper remarked, "Truly Hops are King, and in this region 30,000 Queens are waiting on the old fellow!"[17]

The arrangements made between the hop grower and his city help often involved a number of variables. The price paid to the picker depended upon whether room and board were in-

cluded and what the crop would probably bring at market. In 1868, for example, the growers in Sauk County, Wisconsin, struck a bargain with their pickers, promising them fifty cents a box *and* room and board. This was unusually high compensation, but it was thought necessary to draw out an adequate labor force. Growers could afford to pay comfortable wages when the market price of the crop was high.[18] By comparison, in 1878 pickers in the vicinity of Sangerfield and Marshall, New York, received about twenty-five cents per box with board and forty cents without. In 1893 James F. Clark of Cooperstown was paying his pickers forty cents per box with board and sixty-five cents without.[19]

Frequently, the pickers were accommodated in the farm house. Married couples and single women made their temporary home with the grower and his family, while the men made do in an outbuilding or nearby tenant house. Generally a hop farmer fed his workers at least a noon meal; city pickers took all three meals at the long tables set up for the purpose.

Most pickers were from the laboring class, nearly all of them earning money to support themselves. Some of the men might be idle iron workers, or unemployed carpenters or painters, while among the women were a number of farmers' wives, factory workers, mill hands, and domestic servants. Many were hard-working members of recent emigrant groups. In one field north of Waterville the pickers were from Syracuse, almost all Irish and Germans, and in another nearby yard seventy-five more Syracuse pickers assembled, nearly all Irish. By the late nineteenth century a considerable number of Italians entered the upstate fields. In Wisconsin one wag noted that the pickers were

> English and Irish, Dutch and Danish,
> German, Norwegian, French and Spanish,
> Swiss, Creole, Kanuck and Squaw,
> And every other human you ever saw![20]

On the West Coast the labor pool was drawn from both urban and rural locations, for the shortage of hands was much greater. The distinguishing characteristic of the Pacific fields was that relatively few pickers were of European heritage.

In California the early growers soon realized that native Americans provided the most logical work force, and as a result the business of hop picking in the Far West developed a much

TINGED WITH GOLD

different character. White men were rarely involved in picking the crop, as they often engaged in mining and commerce. Even those who agreed to pull poles were quick to seize a better opportunity elsewhere. On the Pacific Coast there were few women. Single young white women, in particular, often found themselves at a marked disadvantage as pickers, for they often became the center of attention. Thus native Americans were the first to work in the West Coast hop fields. In fact, failing an adequate number of whites, native Americans were invariably preferred by growers, for though they were slow, they were trustworthy, and picked cleanly.[21] Their simple huts of willow brush and crude tents were a welcome sight to farmers searching for additional labor.

In contrast to the states in the Pacific Northwest, however, California did not enjoy the assistance of native Americans for long. In the face of the increasing demand for land, native Americans soon disappeared as a significant segment of the labor force. On the other hand, Chinese were becoming more numerous, particularly in the Sacramento Valley.[22] Towns like Marysville, which boasted the second largest Chinese community outside of San Francisco, became regional centers of field labor.[23] In the mid-1870s experiments using Chinese laborers were painted in glowing terms. Growers believed that the Chinese picked almost twice as fast as native Americans, and speed in harvesting was essential. Cost was also a consideration. As Daniel Flint wrote, without the help of the Chinese picker "hop culture would be unremunerative on account of the high wages demanded by white labor in his absence."[24] In fact, California hop growers generally preferred the Chinese, for they were willing to pick for considerably less. In 1891, on the plantations in the Ukiah Valley, Mendocino County, California, for example, the Chinese help received $.90 per hundred pounds, the native Americans $1.00, and the whites $1.10.[25]

The economic threat this posed to transient whites soon became evident. Whereas the historical record is clear of any opposition to native Americans, by the mid-1870s the general outcry against the "Chinese invasion" was causing serious discussion in the region. White field hands repeatedly complained to the press. Opinionated speeches and impassioned letters were followed. "It isn't his 'cheap labor' only that makes him so objectionable," Daniel Flint wrote, "but his contentment with being a cheap laborer."[26] Despite repeated warnings by more even-tempered civic leaders and law-enforcement officials, belligerent, semimilitary anti-Chinese clubs sprang up in a number of locations throughout the state. Growers soon found that to take advantage of inexpensive Chinese labor invited trouble and possibly more expense, chiefly because of retribution from their own brethren.[27]

One of the earliest indications of the difficulties that would arise during the hop harvest throughout the Pacific Coast occurred near Anaheim, California, in 1877. People in the settlement of Gospel Swamp sent word to J. B. Raines that no Chinese laborers would be allowed in his hop field for the harvest. In mid-August a "Combination of Workingmen of Los Angeles County" published in the *Anaheim Review* a letter addressed to "Vineyard Men and Employers of Chinese Generally," which also threatened trouble. By harvest time the white settlers promised to furnish Raines with all the white men he desired and he agreed that only if the number was insufficient would enough Chinese be employed to gather the crop. Sixty white men signed an agreement to work for Raines, but only a fraction of that number showed up, and those who did called a strike for higher wages. Raines countered by filling their places with native Americans and Chinese. However, when he subsequently dismissed four white men, trouble followed. Upon leaving, one of them threw a potato and struck a Chinese man; the grower drew his pistols and fired. No one was hurt, but because he feared further conflict Raines dismissed all the Chinese and replaced them with native American pickers.[28]

Although federal legislation was passed in 1882 to keep Chinese from entering the United States, their numbers increased; they seemed to be doing "their best to subdue this part of the earth." During the mid-1880s, sentiment against their presence was evident in several hop-growing centers on the Pacific Coast. In Wheatland, California, the Farm Club joined the Anti-Chinese Association to boycott laundries and vegetable peddlers in 1885 and 1886. In 1889 the prominent Sacramento Valley grower S. D. Wood advertised that he needed four hundred to five hundred white people to pick hops, but the notice read, "No Chinese Wanted."[29] As late as 1907, many Sacramento County growers posted notices discouraging both Chinese and Japanese from applying because they intended to supplement white pickers with native Americans.[30] Such job

TINGED WITH GOLD

discrimination was commonplace, even though Asians were willing to work for less.

Naturally, just as the white and native American pickers resented the Chinese, the reverse was also true. As quickly as the Chinese gained a corner on the labor market in a particular region, they attempted to press, en masse, their case for higher wages. In California the "Coolie labor" was particularly well organized and successful in the Sonoma Valley.[31]

The discrimination did not end; instead, the mix of people who worked in the fields changed. In 1891 Daniel Flint wrote: "The pickers range in nationality in the order named: Chinese, Indians, whites, and Japs. As the Exclusion Act crowds out the Chinese, the Japs seem to come in and take their places." He further noted that "white men, women, and children make desirable pickers, and every year sees more of them in our hop yards."[32] By the mid-1890s the whites had almost completely replaced the Asians in most of California's fields. Agents were sent to the mushrooming towns and cities in the state to enlist white pickers, and appeals to tribes in the mountains attracted a few more native Americans, some from considerable distances. For example, although the Piutes from Nevada were finding it increasingly difficult to make the journey as their practice of riding the railroads free was being halted, "Diggers" from Butte, Shasta, and Tehama counties were frequently employed.[33] At the same time, white pickers who arrived in Wheatland came from as far south as Stockton, and as far west as Sonoma County. Writers in the popular literary magazines of the day urged women to participate. "Take a vacation among the hop fields in the gilded early autumn of California," wrote one author. "Your days will be made up of dew-exhaled mornings dwindling to the golden point of noon, of afternoons losing their superfluous heat in sunsets flaming in evening summits, and of nights so cordial and sleep-inviting they seem but moments of oblivion."[34]

In California, where large companies began to dominate the hop-growing industry sooner than elsewhere, advertising was a necessary first step to draw an adequate labor pool to an otherwise isolated location. Advertisements in regional newspapers and posted notices were widely used. In contrast to the poetic prose of the parlor magazines, the facts were laid out in straightforward fashion. In July 1898, for example, the Pleasanton Hop Company issued a circular letter in which the superintendent wrote: "We want pickers. Over 300 acres in hops. Good camping grounds, wood and water free. Pickers must have their own blankets and camping outfit, and are advised to bring their own tents, on the score of economy and convenience. It is believed the crop will be ready to pick August 20th."[35]

Accordingly, dozens of people would begin to migrate toward the fields. Many were the "honest poor" of San Francisco, who would move on to help in the vineyards, and in the apple, pear, and nut orchards in the region. A few tramps strolled in just before the harvest, but they would not remain for long. On the other hand, the equal number of Mexicans that appeared often did remain, a portent of their increasing number after the turn of the century. On a sizeable ranch, such as that of George Hewelett in Hopland, California, nearly two hundred pickers would be camped under the trees by the third week in August. The vegetation would partially hide what could only be described as a tent city.

Whereas in California the smaller hop growers were being eclipsed by larger companies by 1900, in Oregon the cultivation of small hop fields continued alongside the management of large ranches well past World War I. The Willamette Valley provided a variety of working conditions for white, native American, Chinese, and Japanese pickers, and after the turn of the century Bohemians, Italians, and Slavs added to the mix.[36] Residents of nearby cities, villages, and towns, from Portland, Salem, Eugene, Creswell, and Goshen, might flock to the hop fields in Springfield, for example, making use of the steamboats that plied the Willamette River. After the pickers disembarked at the docks next to the hop fields, supplies were unloaded to allow merchants and growers to meet their needs.[37] Meanwhile, native American families would migrate overland with their herds of horses, some from the Grand Ronde and Siletz reservations and others from the Warm Springs Reservation in the eastern part of the state.[38] The Chinese, although they suffered from some discrimination, were never as numerous in Oregon as they were in California and thus played a smaller part in the hop fields.[39]

Some idea of the various appeals made to potential pickers in Oregon is suggested by an advertisement that appeared in a Portland newspaper in the summer of 1907: "Wanted—1,000 hop pickers to pick 624 acres of hops; big crop; largest and best-equipped hop yard in Oregon; all on trellis vine; perfect accommodations; grocery store, bakery, butcher shop, barber shop, dancing pavilion 50 × 150 feet, telephone, physician, beautiful camping ground; 3 acre bathing pool, restaurant, provisions sold at Portland prices. We pay $1.10 per 100 pound; reduced excursion rates on our special train."[40] On the other hand, the contemporary approach of a country newspaper was to appeal to a different class of pickers: "Wanted—1,000 pickers for . . . Hop Field. We pay $1.10 for 100 pounds. Perfect accommodations, good food at city prices, free whisky, dance five nights in the week, evangelists on Sunday and a hell of a good time."[41]

In an attempt to learn exactly who responded to these advertisements and why they came to the fields, sociologist Annie Marion MacLean took a job as a hop picker and surveyed twenty-seven of her coworkers in a Willamette Valley field during the summer of 1907. She found that seventeen were Americans, eight were of German nationality, and two were Swedes. The oldest was fifty, the youngest was fifteen, and the median age for the group was twenty-four. Twenty of the women lived in Portland, Astoria, or Salem; only one came from Washington State. The reasons for working in the fields fell into three categories. Thirteen came to have a good time, "have an outing," or "meet nice men"; eleven made the trip to earn money; and three were picking for their health.[42]

The distinctive character of the hop harvest in the Washington Territory was largely attributable to the native Americans. Only a small proportion of the pickers were white, and even fewer were Asian. In 1876, for example, about fifteen hundred native Americans were employed in the Puyallup Valley, supplemented by about two hundred Chinese.[43] With the tremendous expansion of the fields in that area, apprehension ran high among growers the following year because a preharvest tally indicated that nearly three thousand workers were needed. The growers had little recourse but to recruit the native Americans en masse. They responded enthusiastically. In fact, their appearance in 1878 was so startling that a military company of sixty-four men was organized at Sumner, in Pierce County, for the purpose of maintaining order during the harvest season.[44]

The Puget Sound area became the terminus for thousands of native Americans every year at harvest time. From their homes in British Columbia and even Alaska, scores would paddle and sail down the coast in long canoes fashioned from im-

4.4 Native American
canoes and camp, Ballast
Island, Seattle, Washing-
ton, c. 1882. Long, colorful canoes filled the water-
front. (Washington State
Historical Society,
Tacoma, Wash.)

mense cedars. The Siwashes of the Washington coast joined
the Chimsyans and Haidas of British Columbia, and the Thlin-
kets of southeastern Alaska.[45] In certain localities, along the
wharves of Seattle for example, an unoccupied piece of land in
the commercial section of the city would harbor a fleet of native
American craft. Others came overland from distant reservations, traveling as families and tribes, with colorful woven
baskets, blankets, mats, rugs, and perhaps a herd of horses to
sell.[46]

In Puyallup, the principal early marketing center, the mix of
native Americans, whites, and the few Chinese created a tour-
ist attraction in the Northwest. Puyallup streets were crowded
for two or three weeks every autumn with visitors, who drove
in carriages from Seattle, Tacoma, and other places, eager to

TINGED WITH GOLD

see the hop pickers at work and to visit the various native American camps in the vicinity. Around the turn of the century, the street cars were loaded to their capacity going out to the fields, and they were even more crowded on the return trip, for few observers left without purchasing one or more specimens of native American basketry.[47]

The accommodations occupied by the native Americans varied widely. On one hand, large growers erected tenant housing able to shelter a limited number of pickers. For example, the Snoqualmie Company, located on the shores of the Willapa River where it meets the Pacific Ocean, erected twenty sheds to house native American pickers in addition to the dozen dwellings occupied year-round by the native American families employed on the premises.[48] On the other hand, many smaller growers merely provided an opportunity to camp. Native American housing was often the most jerry-rigged of any. Rarely was a teepee to be found; but tents and shanties of all kinds were built, in every sort of layout. Long rows of split red salmon gleamed in the sun, hanging to dry, symbols of the freedom of movement associated with these people. The smell of the native American camp was unmistakable.

The favorite pastime of the native Americans was gambling.[49] The presence of other tribes seemed to awaken in them

*4.5  Native American encampment at the west end of Vine, Cedar, and Broad streets, Seattle, Washington, c. 1882. Subsequent development brought wharves, warehouses, shops, and railroad lines. (Washington State Historical Society, Tacoma, Wash.)*

*4.6 Ezra Meeker's hop field and camp, Puyallup, Washington, 1882. The haphazard arrangement of various kinds of tents can be seen in the foreground; the Native Americans are gathered near the corral adjacent to the fields. (Meeker Collection, Washington State Historical Society, Tacoma, Wash.)*

*4.7 Hop culture in the Yakima Valley, Washington Territory, 1887. The stylish homes of the earliest growers are juxtaposed with the Native American pickers in the fields. (From* West Shore *13, no. 10 [October 1887], frontispiece.)*

the desire for a contest. The Swinomish might be pitted against the Lummi, for example, and the struggle would go on into the night with the repeated throw of the dice causing at a minimum wailing and chanting, while money, livestock, clothes, and provisions changed hands. In addition, a race track, laid out where North Puyallup was built, became known as the "Devil's Playground" and contributed more than its share of the excitement to the area. Crowds swarming near the horse corrals moved to and fro around the grounds.

The growers in the Yakima Valley also depended almost completely upon native American labor. Yakima, a small, isolated inland town until the Northern Pacific relocated it in 1886, quickly mushroomed thereafter. What the railroad did for the city by providing for irrigation and sanitation improvements, the Moxee Company did for the countryside by advancing the growth of tobacco, corn, and cotton. Hops, however, were most important and have continued to this day as a primary crop.

The fields in Washington appear to have been some of the most highly segregated, with whites in one area of the field, native Americans in another, and Chinese, when there were any, in a third. The whites were the most highly regimented. By contrast, the native Americans were the least formal: members of a family would group around a single box, mat, blanket, or basket. The mothers, some with papooses strapped to their

*4.8   Hop pickers at an unidentified farm, McMillan, Washington, c. 1885. A company of white pickers with several children and young adults. This is before the introduction of the trellis system to the Northwest. (Washington State Historical Society, Tacoma, Wash.)*

4.9 *Native American pickers, unidentified farm, White River Valley, 1902. (Washington State Historical Society, Tacoma, Wash.)*

4.10 *Native American encampment, unidentified farm, White River Valley, 1902. The use of straw mats to create living quarters was a commonplace solution to providing temporary housing used by Native Americans in the Puget Sound area. (Washington State Historical Society, Tacoma, Wash.)*

backs, would pick assiduously, while the old patriarch would stand by, planning how to approach the next vine.

The relationship of the Chinese to the white majority in the region was often tense and, as in California, at times violent. Tacoma sprang into the national limelight in 1885 when its citizens forced the city council to fire all Chinese employed by the city. Not satisfied with this measure, the whites tried economic sanctions and boycotts, and a few even stoned and torched the houses of Chinese, without fear of punishment.[50]

Another serious conflict occurred in the fall of 1885, when the Wold brothers, of Squak Valley, Washington, hired a number of Chinese laborers. On the afternoon of September 5 about forty Chinese pitched their tents on the Wold farm; that night an angry white mob threatened them, but the encounter stopped short of violence. The next day an armed group of white men turned back a second party of Chinese. On September 7 at about 10 P.M., five white men and two Indians began firing rifles and pistols into the group of tents, in which thirty-seven Chinese were asleep. Three were killed, and three others were wounded; the next day the Chinese left. Although the identity of the participants was known and they were indicted, no con-

TINGED WITH GOLD

victions were ever secured.[51] As in California, in a few years the Chinese were completely superseded by white labor in the hop fields, and difficulties subsided.

For many city pickers in the West, as in the East, the procedure for travel was rather simple: one needed only to secure a ticket from an agent and show up at eight in the morning to board the "Hop Special." However, some fields were not very accessible by rail. In addition, pickers who lived in rural areas often were unable to get to a railroad. In these cases, travel was more difficult and time-consuming. The arrival of hop pickers in Independence, Oregon, for example, was for some the end of a pilgrimage of hundreds of miles.

White pickers from rural areas often arrived in a big frame hay wagon loaded with bedding, furniture, and cooking utensils. After being unloaded, the wagon was parked outside the tent, while the family cow, which had so faithfully followed, wandered unrestrained around the canvas home.[52] With the advent of the automobile and more widespread use of refrigeration, the hardship of travel was not as great and milk was more generally available, but the distances were little shortened and the baggage never lightened.

*4.11 Hop pickers arriving at Independence, Oregon, c. 1920(?). Teamsters await at the station as field hands disembark. (Lane County Historical Museum, Eugene, Oreg.)*

Having arrived in the hop-picking region, the field workers made their living accommodations as pleasant as possible. As in the East, if the hop field was a modest one, women could expect to live in the farmhouse, while the men would occupy the barn, ranch house, or another auxiliary building fitted up for the purpose. Three or four rope beds with cornhusk mattresses, sheets, and blankets, with perhaps a stool or well-worn chair, were sufficient furniture. Alternatively, tents would be pitched alongside the vegetable garden or next to the hop fields.

On large farms or ranches, pickers found dormitories or tenant houses, and sometimes commissaries and dining halls. Simple frame "apartments," with a door in front and a window in the rear, would serve as temporary homes for many families. Small children, often left behind in camp with an elderly relative or an older sibling "in charge," found their activities bounded by the walls of these structures. Meanwhile, single pickers either rented tents from the grower or brought their own.

If provisions were needed, a delegation of pickers would make a trip to a nearby town to visit the grocery, clothing, and hardware stores. Hop-picking supplies included the obvious necessities like gloves and bonnets, to protect the hands from being cut and shade the face from the bright sun. City pickers and others who arrived by rail purchased housekeeping necessities locally. Cookstoves, scrub brushes, and groceries, especially meat and produce, were in high demand.[53]

*4.12  Hop pickers' tents, western Washington, c. 1900. Camping was a way of life during the harvest season. (Washington State Historical Society, Tacoma, Wash.)*

TINGED WITH GOLD

In town, even if the store owner had hired additional clerks for the harvest season, most pickers found long lines outside commercial establishments as trade doubled, and doubled again. Delivery wagons and carts groaned under the weight of sacks of flour as they made their way slowly toward the camps. For pickers who did not have their own transportation, a small store on the grower's grounds generally provided the necessary staples. Further, the picker was probably safe in assuming that a traveling merchant or produce vendor would visit the fields during the week.

The work of food preparation and cooking was followed by cleaning in what seemed like a never-ending cycle. In smaller fields the pickers often shared these tasks with the grower's family. In larger fields the dining-hall staff would take charge.

*4.13  Hop pickers outside their temporary living quarters, unidentified farm, Oregon, 1914. These shed-roofed buildings served as temporary housing but offered little privacy. (Oregon Historical Society, Portland, Oreg.)*

Meals were served in the early morning, near mid-day, and around 6 P.M., when scores of pickers sat down at long tables next to the kitchen. Unsuspecting young women, promised a free meal if they would help out the short-handed staff by waiting on tables in the dining hall, might soon find that they were being pressed into service washing stacks of dirty dishes rather than taking part in the harvest. As with much of the food preparation, clothes washing often took place out of doors. This was another major undertaking; wash basins and pails were in great demand, and clotheslines were improvised everywhere. Of course, wood piles, wells, and water barrels were frequently the meeting places where romance would begin.

In the large western fields pickers often arrived during the weekend. After camp was made, they would relax and get to know their temporary neighbors. On Sunday the camp evidenced a democratic character. Everyone, regardless of origin or background, was more open and tolerant. In the evening the smells of a wood stove cooking a late supper of fried bacon and onions would combine with the aroma of fresh hay, used for bedding, while oil lamps—later replaced by bare electric bulbs—would provide pale illumination. A spritely hop or an evangelical preacher would be the evening's entertainment.

The first contact many pickers had with the grower's staff was on Sunday afternoon or Monday morning, during registration. Early in the day a long line began to assemble. A badge or perhaps a ribbon with an identification number issued to each picker indicated the yard and division or company to which he or she was assigned, while the yard bosses would write down the number, the name, and perhaps a few basic items of identification for each person. With registration the picker's anticipation of the harvest peaked.

Work in the fields generally began on a Monday morning, although growers followed no set rule. In a small yard the ambitious picker might enter the yard at the crack of dawn and pick a box or two before breakfast. On larger farms and plantations, companies would form and march off to the fields in a military manner. When reaching a division, everyone chose a partner so as to pick two to a hill. Pickers then equipped themselves with baskets and a canvas bag into which to empty the baskets.

When pickers began work, they were assigned four to a box, with one additional person—the box tender—assigned to pull

the poles. In large yards, each group was called a section. Eighteen sections constituted a division, and nine divisions made a yard. Thus, the grower knew every picker by his or her place in the yard.[54]

The divisions were distributed throughout the fields, each day picking so as to draw nearer to one another. Men driving teams drew the hops from all sections at once and divided them among the kilns, so that a portion of each division's crop was distributed to each kiln. This guaranteed consistency in every bale of hops.

To the pickers, the pole pullers and the wire men were the most important figures in the field. These men, who were responsible for letting down the hops, controlled the pickers' rate of progress. A group of pickers who wished more hops would call out "Pole man!" or "Wire down!" and listen for heavy tramping over the rough-plowed soil coming in their direction. The pullers or wire men were the aristocrats of the company. They earned better wages and possessed considerable prestige; currying their favor meant a quick response, a fuller box, and more money.

In the East, even as late as the mid-nineteenth century, pickers were paid in a number of ways. In Otsego County, New York, for example, in 1853, women picked by the week, the box, and the bushel in the same field. As one grower wrote, "The latter [by the bushel] presents itself to me as altogether

*4.14 Hop pickers at Louis Grouter's yard, Allerton, Washington, 1911. Considerable ethnic diversity is evident in this field, which is making use of the trellis system. The wire men hold their cutting poles. (Washington State Historical Society, Tacoma, Wash.)*

*4.15  Hop box tickets, J. H. Davis and R. B. Brown farms, Oneida County, New York, c. 1890. Note the crispness of the corners and edges of the ticket worth five boxes as opposed to the well-worn examples that were worth less but used more. (Courtesy of John Alden Haight.)*

preferable. Then we pay those who earn their money, and those who do not surely ought not to have it."[55] In fact, picking by the box became the most widespread method of payment. In exchange for one or more boxes of hops, the picker collected tickets of various colors and denominations, and the field supervisor recorded the credit in a small notebook. For example, pickers on the farm of R. B. Brown, of Oneida County, New York, were issued a red ticket, which signified five boxes; a golden yellow ticket for one box; or a green ticket, for half a box. These were redeemed weekly or at the conclusion of the harvest.

On the West Coast, however, boxes were not the only containers used to collect the harvest; native Americans regularly brought baskets. Particularly on larger farms on the Pacific Coast, the pickers came to be paid by the pound, and thus it was necessary to weigh each person's hops. In the early twentieth century a more widespread method called for a weigher to make out duplicate tags, indicating the picker's number, the number of sacks, and the weight of the hops. The weigher gave one tag to the picker and filed the other in the office. When the pickers came to receive their pay, they presented their tags, and the weights were added and checked with the duplicates that had been collected in the field.[56]

On some farms and ranches hop tickets could be redeemed weekly. Those pickers who did not wish to stay long in the fields were especially careful to investigate this point before setting up camp. Other pickers simply wanted the cash to quench their thirst for beer or liquor. Local tavern owners and merchants often considered the tickets negotiable tender, so that those who wanted to get drunk could.

Pickers also found that on some farms and ranches there was only one payday, at the end of the harvest, and growers explicitly stated the penalties for leaving early. As soon as James Clark's employees arrived by railroad, they were marched in details to the boarding houses, during which time their attention was drawn to notices posted throughout the camp, which they were expected to read, or have read to them, and understand. By going to work in Clark's fields, the picker agreed that if he or she left the yards before the end of the harvest, for any cause whatever, he or she would be paid ten cents per box less than the ruling price. Box tenders who left early received twenty-five cents less. Indeed, to ensure that Clark had no trouble with his employees, he made sure that the special train on which his pickers left Cooperstown was well out of town before they were paid anything at all.[57]

While some pickers might want to be paid almost immediately, others would be patient and wait. The northern native American pickers employed in the Puyallup Valley in 1890 actually preferred to wait until the end of the season to get their tickets cashed. This suggested that the native Americans intended to take their money with them, much to the consternation of the local merchants.

Although one of the first objectives of the picker in going hop picking was to make money, another was simply to have fun. After pickers had put in a full morning and long afternoon in the field, their more immediate thoughts turned to a solid supper and the evening's entertainment. While pickers enjoyed a considerable amount of singing and joking in the fields, the social highlights of the hop-picking season were the dances, called "hops."

In the early nineteenth century the proprietor often celebrated the harvest with a public ball in a nearby village, inviting everyone involved. Matthew Vassar, a brewer and the founder of Vassar College, was many times a spectator at the hop balls at the American Hotel in Waterville. The great Assembly Room there would be filled with pickers from a number of fields, listening to the strains of a melodeon band.[58] Many smaller inns and hotels held dances on Tuesday, Thursday, and

4.16 *Hop pickers were encouraged to order their hop tickets early. The tickets were filled in by the weigher and punched in the field. (From* Pacific Hop Grower *6, no. 4 [August 1936]: 5.)*

Saturday, and on alternate nights different farmers would put on dances in their barns.

In the late nineteenth century, as animosity against the city pickers rose in many upstate New York communities, the entertainment of the pickers was more likely to be restricted to the farm. James F. Clark built a public hall that accommodated one hundred people and sponsored dances once a week. The Albany pickers provided a string band.

In the Far West white workers often got together during the evening to gossip, sing, and occasionally enjoy the music of a violin or accordion.[59] A Saturday evening in August on a California hop ranch would normally include a square dance in the cooling barn. Just before the stroke of midnight, the music stopped, the benches and chairs were removed, the lamps were extinguished, and everyone scrambled out to make way for a freshly cured load of hops to be dumped.[60] Asian camps were relatively quiet by comparison. Chinese pickers, who preferred quiet recreation, chose to smoke and sleep at night.

In stark contrast to the colorful prose of writers in the popular press and promoters describing the beauty of the fields, the lightness of the work, and the camaraderie of the workers were a number of rumors about exploitative growers and poor field conditions. The Wheatland hop-fields riot of August 3, 1913, was perhaps the most infamous example of labor difficulty in California's inner valley up to that time. To understand how hop growing in California had graduated from a commercial enterprise to an agricultural industry, it is essential to bring this story to light. All the economic and social difficulties associated with the struggles between capital and labor that could be found in other agricultural industries were evident by this time in the golden fields of Yuba County.

The scene of the dispute was the well-known Durst hop ranch at Wheatland, where at least two thousand pickers, wire men, haulers, and drying men found employment every season. Advertisements, which had been broadcast throughout California, Nevada, and southern Oregon, stated that the picking season would begin about August 1, so that by July 30 a fairly large crowd had assembled. Durst invited difficulty when he advertised that all white pickers who made an application before August 1 would be given work. This offer was impossible to fulfill, for many more applied than could be employed.[61] Further, evidence introduced at the subsequent trial indicated that

TINGED WITH GOLD

from about one-third to one-half of the pickers were foreigners, including Hindus, Italians, Japanese, Mexicans, Puerto Ricans, Poles, and Syrians. A field boss counted twenty-seven nationalities in his picking gang of 235 people. The Americans were chiefly dirt farmers, occasional miners, ranch hands, hoboes, and a number of families who were in the habit of using the hop and fruit seasons as their "country vacation."[62]

At the time of the outbreak an estimated twenty-eight hundred pickers were camped on a low, unshaded hill. The management of the ranch was poorly prepared to provide either housing or sanitation facilities for such numbers. Durst rented tents to field workers for seventy-five cents per week, but all of these were soon taken. Some old tents were donated free of charge. Aside from these, only "bull pens," or gunny sacks stretched over fence rails, provided a roof over some of the pickers' heads. Before these accommodations were ready, many slept in the open fields, some on piles of straw. The water supply was insufficient. Two of the wells were dry by sun-up, forcing pickers to walk to more distant wells on the ranch or even to town for water. No drinking water was furnished in the fields, despite heat that averaged 105 or 106 degrees in the shade. Water had to be carried out by the pickers in bottles and jugs, and this became warm and unpalatable in an hour or two. Durst forbade

*4.17  View of the hop driers and trellised fields, Durst Hop Ranch, Wheatland, Yuba County, California, 1913. The broad, flat valley was virtually uninterrupted by tree lines.*

*4.18 The Wheatland hop fields, August 1913. Closeup of the Durst campgrounds before the violence erupted. (Archives of Labor and Urban Affairs, Wayne State University, Detroit, Mich.)*

local Wheatland stores to send delivery wagons to the camp. This meant that pickers had only one immediate source of supplies and provisions, a grocery concession located in the center of the camp grounds. Pickers later discovered that half of the net profits earned by this allegedly independent grocery store went to Durst.[63] A concession to sell stew had been let, and a stew wagon went out about lunch time among the pickers. If an employee purchased some stew, he or she could get a glass of water with it. Later investigation disclosed that the ranch owner's cousin, Jim Durst, obtained a commission to sell lemonade to the field workers for five cents a glass. However, the druggist from whom he purchased his supplies testified that the lemonade was made entirely from citric acid![64]

The management had made no provision for the disposal of garbage. The trash and refuse that collected became not only a source of pollution, draining into the water supply, but also the

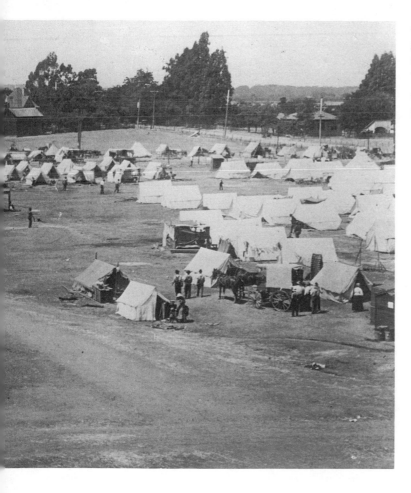

4.19 *View of the camp-grounds from the water-tower, Durst Hop Ranch, Wheatland, Yuba County, California, 1913. Scene of the famous riots.*

4.20 The Wheatland hop
fields, August 1913. The
women and children who
worked on the Durst ranch
at the outset of the harvest.
(Archives of Labor and
Urban Affairs, Wayne
State University, Detroit,
Mich.)

feeding place of millions of flies, and the garbage caused an unbearable stench throughout the camp.

The most disgusting sanitary abuse was the condition of the toilets. For twenty-eight hundred people Durst provided only eleven toilets, each accommodating two people. Often lines of fifteen to twenty men, women, and children waited to use them. The toilets were crude boxes placed over holes in the ground about two feet in depth. By the end of the second day the toilets so reeked of their semiliquid mass of filth that those who merely passed by them suffered from nausea and vomiting. Many of the campers soon refused to get near the "facilities" and resorted to using the fields among the vines.[65]

All this was not uncommon among similar ranches. The general discontent and camp-fire grumbling needed a further reason to crystallize into a plan of action, and that was found in the wage system presented to the pickers. The "going rate" for hop picking in California at this time was one dollar per hundred pounds of hops. For each hundred pounds the Durst management, however, was giving out a check that could be cashed in for only ninety cents, with a ten-cent "bonus" added

if the picker stayed through the entire harvest. The bonus, it seemed, was a holdback from the going wage and allowed the employer to benefit if the picker left before the end of the season.

Clearly Durst planned to bring more pickers to his fields than he could possibly keep employed. The hop-drying kilns of the ranch could not cure the amount of hops produced by more than fifteen hundred pickers. Nearly half of the campers merely hung around, waiting to be admitted to the fields.

Furthermore, the pickers claimed that Durst demanded an unusually high standard of cleanliness for his hops. With hundreds of campers anxiously waiting at the gate, inspectors held an additional psychological weapon against those already in the field. The inspectors often forced pickers to walk to the kilns and pick their bags over, which kept the average day's earnings intolerably low, below ninety cents per day.

At a meeting of the campers on Saturday evening, August 2, a number of laborers vented their resentment. In the summer twilight, a number of speakers denounced the wages and living conditions. Durst, who happened by in his automobile, was

*4.21 The Wheatland hop fields, August 1913. Hop pickers meeting to discuss the turn of events. (Archives of Labor and Urban Affairs, Wayne State University, Detroit, Mich.)*

4.22 *The Wheatland hop fields, August 1913. Agitation among the hop pickers. The majority of those assembled were men. (Archives of Labor and Urban Affairs, Wayne State University, Detroit, Mich.)*

hissed. He refused to deal with the crowd but asked that a committee visit him the next morning with a formal list of complaints.

At ten o'clock the next morning, a committee headed by "Blackie" Ford presented Durst with a list of demands, aimed at improving living conditions and raising the wages of the pickers to $1.25 per hour. Durst agreed to try to accomplish the former, but refused to change the pay scale. Shortly thereafter, in front of a crowd, Durst ordered Ford to leave the ranch and flicked him in the face with a glove. At Durst's insistence a deputy officer named Reardon, from Wheatland, later attempted to arrest Ford as a trespasser, but Ford demanded to be served with a warrant, which the officer did not have. As people jostled the deputy he drew a revolver to force back the crowd, whereupon an excitable Italian jumped in front of a girl and bared his chest crying "Don't shoot a girl—shoot me!" This encounter further excited the crowd.

Throughout the afternoon emotions approached a fever pitch as "Blackie" Ford, a former member of the Industrial Workers of the World, and about a hundred other union "card men" urged the crowd to "stick together."[66] At a general meeting at two o'clock on Sunday afternoon, IWW songs were sung and speakers openly challenged Durst with a strike. The climax of

this drama occurred as Ford, having taken a sick baby from its mother's arms, held the child before the fifteen hundred people who had assembled and cried out, "It's for the kids we're doing this!" At that point the sheriff's posse arrived in several automobiles. District Attorney E. T. Manwell, Sheriff George Voss, and Deputy Eugene Reardon made their way to the front of the crowd, calling for the people to disperse. Reardon, spotting Ford, said, "There's your man, sheriff. I have a warrant this time." However, as the officers closed in, a bench on which some of the crowd had been standing crashed down, causing a deputy at the edge of the group to fire in the air "to intimidate them," as he later explained. A volley of shots followed, and for a few minutes pandemonium reigned. Part of the posse ran back to their automobiles and escaped; part of the crowd chased a second deputy into the ranch store, where he managed to barricade himself. When the crowd scattered, the dead bodies of the district attorney, the deputy, and two unknown pickers lay on the ground next to the sheriff, who was unconscious.[67]

All night frightened campers fled the ranch, so that the following morning, when the state militia arrived, calm had already been restored. During the following weeks law-enforcement officials searched for the ringleaders of the crowd and for those directly responsible for the shooting. The aggressiveness with which the authorities pursued the case led to several false arrests, and as a result the pickers organized a committee to raise funds for the defense of prisoners.

Finally, in January 1914, four men, including Ford, were brought to trial and charged with the murder of District Attorney Manwell. The court convicted two of these men, including Ford, of second degree murder on the grounds that they had inspired the crowd to violence. In fact, there was no testimony that Ford was seen with a gun or used one. On February 5, 1914, the judge sentenced both men to life terms in Folsom Prison, and their convictions were sustained upon appeal. Ford was paroled twelve years later.[68]

As a result of the Wheatland difficulties, Governor Hiram Johnson appointed the Commission of Immigration and Housing to investigate ways in which the living and working conditions of migrant farm workers could be improved. Two paths were obvious: upgrade the facilities and provide a number of social services. With the advent of World War I and Prohibition, however, the recommendations were not acted upon for a few years. By that time California was no longer a major hop-growing state; the fields of Oregon were claiming first-rank importance. There the efforts to appease the pickers would continue.

In 1922 the Independence, Oregon, Chamber of Commerce undertook to centralize the entertainment facilities for hop pickers. The ostensible purpose was to improve the relationship between the pickers and the community, but it was also a thinly disguised excuse for bringing hop pickers' money to town. Going a step further, in 1927 the chamber staged a street carnival, which was the first of a number of hop festivals held annually until World War II. To celebrate the harvest, a hop queen and her court were elected by the merchants, who supported a parade with marching bands and floats, followed by dances. A rodeo might take place the last two nights of this three-day event, which also featured wrestling matches and games.[69]

Effective improvements were seen in some fields. On the Eola ranch of the E. Clemens Horst Company, J. C. Henderson, community service executive and supervisor of recreation for the city of Portland, began a popular social initiative. The three-pronged program included first, housekeeping, tent pitching, sanitation, water, and fuel supply; second, primarily

TINGED WITH GOLD

providing first aid and preventative medicine but also issuing a
camp newspaper and supervising concessions; and third, a recreation program that included the installation of playgrounds
and nurseries and assumed the responsibility for nightly camp
fire meetings, dances, religious services, and special programs
on Sunday afternoons. Not only were sandboxes, swings, and
slides built under huge pine trees, but a playground worker and
a nurse supervised over one hundred children in daily activities. Other large ranches throughout the region adopted many
of these activities in one form or another.[70]

In 1937 an investigation of the hop camps in the Yakima
Valley indicated that housing conditions varied, but on the
whole they were tolerable.[71] Two-fifths of the heads of house-
holds who were living on the farms where they were employed
found shelter in permanent cabins. The remainder were dwelling in tents. About one-sixth of the pickers were crowded,
living in shelters that housed six or more persons. On the other

*4.23   Unidentified hop
ranch, Lower Yakima Val-
ley, Washington, c. 1938.
The accommodations for a
seasonal labor force of
1,500 workers. The cabins
housed migrant families;
the tents sheltered single
itinerants and Native
Americans. (Washington
State Historical Society,
Tacoma, Wash.)*

hand, about half of the pickers were dwelling in shelters in which only two or three pickers were living. The sanitary conditions, as judged by the county board of health, continued to be inadequate. Both water supply and drainage were insufficient. The general health of the pickers was also poor. Inspectors found illnesses, particularly dysentery, in nearly all camps, and medical treatment was almost nonexistent. Growers paid considerable attention to providing recreation facilities, however, by supporting an open-air theater, dances, beer parties, and group musical events. Native American pickers who gambled were not kept from doing so. Remarkably, some surveys indicated that almost half of the total number of pickers had worked in the Yakima Valley previously, and it was likely that many would return. Unfortunately, soon they would be no longer needed: the mechanization of the hop fields was advancing ever forward.

The principal concerns of the picker were transportation, living accommodations, working conditions in the field, and wages. The smart hop picker knew the importance of gaining the best information about these specifics. Although there were problems between the growers and the pickers, the differences between various groups of pickers were often of greater concern. The fields attracted a wide range of society. Whether home pickers or city pickers, Chinese or native American, the ethnic mix reflected the regions from which they were drawn. Regardless of ethnic background, however, women were responsible for doing the majority of the picking, largely because they were capable and available. This gender imbalance in the field remained part of hop culture until the advent of World War II, when women responded to the call for manufacturing jobs. Thereafter, the colorful days of the hop picker all but came to a close.

# ❧ 5 ❧
# Hop Kilns, Hop Houses, and Hop Driers

D rying was the principal means of preserving most farm products. Not surprisingly, then, open-air drying of hops was commonplace for growers throughout the seventeenth, eighteenth, and nineteenth centuries. As late as 1853, the prominent agricultural writer Solon Robinson indicated that hops need only be "spread awhile" to be dried.[1] Although this method may have been adequate for household production, commercial growers could not wait for the sun to assist them. Drying had to be accelerated if hop growing was to be an economically viable enterprise. Growers soon realized that structures for drying also had to facilitate cooling, baling, and storing the crop. Each of these operations involved a number of smaller steps that became the object of increased attention and led, in turn, to various inventions and improvements. The change from the simple banked hop kiln, through the more inclusive hop house, to the hop drier and its attendant cooling structures represents an evolution of almost two hundred years.

Because of the close association between colonial New England and the mother country, a brief review of the arrangements employed in Great Britain is useful as a reference point. There, rectangular single-story frame buildings, called "oast houses," were commonly used to cure hops through the late eighteenth century. A woodcut illustration of an oast house in Kent included in Reynold Scot's *A Perfite Platforme of a Hoppe Garden*, published in 1574, clearly shows the plan of this traditional structure. It was nine feet high, eight feet wide, and eighteen or nineteen feet long, and it contained three rooms: the first was the collecting room, for green hops; in the center was an oast or drying room, eight feet square; and the last was the cooling room. The brick furnace, over which the hops were placed to dry, measured seven feet long and thirty inches high. The bricks were set "checkerwise" in a honeycomb pattern, so that the heat could rise between them, and piled wide at the bottom and narrow at the top, "somewhat like the roof of a house."[2] The drying floor consisted of wooden lath strips, laid a quarter of an inch apart, five feet above the ground level, that is, about two and a half feet above the furnace. In the drying process moisture escaped through the roof or gable openings, while the hops became tinged a golden brown.[3]

This building type continued to be constructed almost without change through the eighteenth century.[4] The more ad-

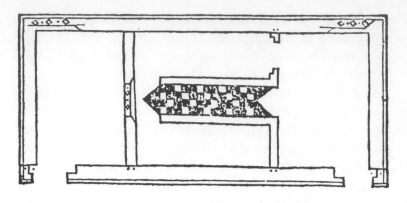

*5.1  Hop house, Kent, England, c. 1574. This plan shows the traditional layout, in the center of which is the oast, or hearth. (From Reynold Scot,* A Perfite Platforme of a Hoppe Garden *[1574; repr., New York: DaCapo Press, 1973], 41.)*

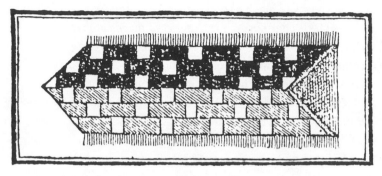

*5.2  Brick furnace, hop house, Kent, England, c. 1574. The checkerboard-like pattern indicates how bricks were placed to allow the heat to rise in the spaces between them. (From Reynold Scot,* A Perfite Platforme of a Hoppe Garden *[1574; repr., New York: DaCapo Press, 1973], 40.)*

vanced methods of controlled drying employed in malt kilns proved a source of new ideas, borrowed for trial in hop oasts. One alternative was a combustion chamber consisting of a brick furnace with a plaster-on-lath flue, widening like an inverted pyramid to meet the drying floor. About 1780 these were superseded by continuously coved brick flues, widening in the same fashion.[5] As production increased, oast houses became larger and more sophisticated. By the 1830s, for example, more than one oast used as many as four drying floors, stacked one above the other.[6] Growers employed charcoal and coke as fuels because they were virtually smoke-free, and sulphur was thrown on the fires to improve the appearance of the hops by bleaching them.

Single-story frame structures of this kind may have been constructed in the American colonies, but no examples have come to light. Most of the sites where the earliest commercial production took place, in Middlesex County, Massachusetts, have been altered. In fact, few pre-Revolutionary agricultural structures of any kind exist outside the Connecticut River valley and southeastern Pennsylvania, and neither of these areas is known for cultivating hops.

On the other hand, the earliest published references to commercial operations in what would become the United States clearly indicate that developments that occurred in Great Britain were of influence across the ocean. Commercial hop growing in this country seems to have been boosted with the introduction of the charcoal-fired kiln. The first hops to be dried in a charcoal-fired kiln were cured in Massachusetts in 1791. A Scottish brewer named Laird visited the hop yards of Samuel Jaques, Jr., seeking hops. Dissatisfied, Laird provided Jaques with the particulars of European drying methods. The experiment proved successful and caused so much excitement in the neighborhood that the next season every hop grower followed the same method. In addition, "some further improvements in the construction of the kilns was adopted."[7]

Hop growers kept abreast of the latest developments abroad not only by contact with a number of foreign-born emigrant experts but also by reading an increasing amount of literature that dealt with agrarian concerns. By the post–Revolutionary War period, English agricultural books and journals were relatively common in the eastern United States, so that those who wished to could gain a clear understanding of hop culture and the manner of building a hop kiln.[8]

A typical reference text of the period was *The New Cyclopaedia; or, Universal Dictionary of Arts, Sciences, and Literature*, by the Englishman Abraham Rees. First published in London beginning in 1802, the English edition of the multivolume work was completed in 1820. An American edition, "revised, corrected, enlarged, and adapted to this country by several literary and scientific characters,"[9] was issued about 1818 in Philadelphia. Rees described an oast as a square room measuring ten feet on a side in the middle of which was a fireplace, or herse, about thirteen inches high and wide, and about eight feet long. By means of several holes in the sides of the herse, the heat of the charcoal fire was "let out" into the room and rose beneath a hop-drying floor made of laths an inch square laid a quarter inch apart, supported by beams and covered by hair cloth. The distance between the fireplace and the drying floor was five feet. The hops, laid on the floor six or seven inches deep, were restrained by walls four feet high until

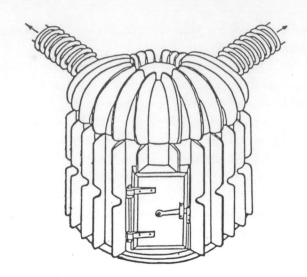

dry, when they were raked through a window into an adjacent room to cool.[10]

In this simplified model, collecting the green hops was not a concern. Once again, three rooms sufficed, although these differ from the example described earlier. The kiln consisted of a fireplace room to generate heat, with the drying area above and the cooling room alongside.

The preference for charcoal as a smokeless fuel is important in light of the widespread availability of wood in the East. In both Great Britain and the United States growers introduced stoves to economize on fuel. The entry in the *Cyclopaedia* went on to state that in England, where charcoal was expensive, "many people have adopted the method of drying with sea-coal, upon what they call cockle-oasts, which are square iron boxes placed upon brick work and a flue and chimney in the back part of the building for the smoke to go off."[11] The cockle oast, or cockle stove, was the earliest cast-iron hop-drying aid. Whether this device was square or cylindrical, growers preferred it for its long-term economy.[12] It should be noted, however, that a cockle oast required a significant initial investment. No evidence has come to light indicating that cockle stoves were imported to the United States.

Although Rees's work did not include an illustration of an oast house, the repeated mention of its similarity to a malt house, which was illustrated and described at length, indicates a transfer of technology between these two structures. The

TINGED WITH GOLD

malt kiln was a two-story structure with a furnace or stove on the first floor and a drying floor above, capped by a low pyramidal roof. The heat source generated hot air that circulated in two stovepipes that radiated heat beneath the drying floor. The moist, warm air was collected as it rose in the domed ceiling and allowed to escape through a ventilator atop the roof. The design of this ventilator was similar to the cowl of an oast house, which took the form of a conical cap, open on one side for about one-third of its circumference. These devices were a logical response to the necessity of protecting the hole in the apex while providing the maximum draft possible in the kiln, for they turned by means of a wind vane. This British invention of the 1790s was perhaps the most notable feature of the more sophisticated malt and hop kilns.[13]

Better known to agriculturalists in the United States than Rees was John Claudius Loudon. A prolific English essayist and compiler, Loudon put forward dozens of articles and books,

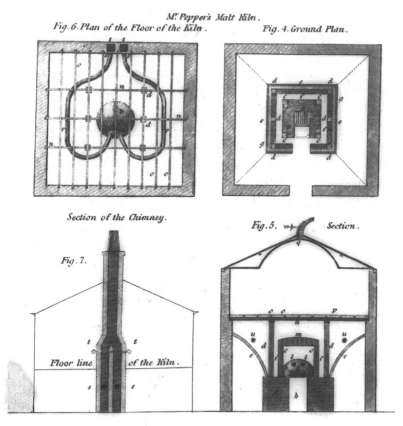

5.4 *Malt Kiln, England, c. 1816. The sophistication of these kilns was a likely source of inspiration for hop kilns. (From Abraham Rees, ed.,* The New Cyclopaedia; or Universal Dictionary of Arts, Sciences, and Literature *[Philadelphia: Samuel Bradford, 1818], n.p.)*

many of which were widely quoted on this side of the Atlantic.[14] In his *Encyclopedia of Cottage, Farm, and Villa Architecture and Furniture*, he presented the more traditional hipped-roof kiln for drying hops, but then he described at length the virtues of the round kilns erected in 1832 at Teston, Kent. In that instance, two pairs of circular brick kilns, twenty feet and sixteen feet in diameter, were connected to a stowage room. The designer, John Read, adopted this plan as early as 1796 because it contained more area than any other figure with the same length of exterior wall, and thus it cost less. Read apparently superintended the erection of hundreds of these kilns, but they received little attention outside of Kent and Sussex.[15]

An excellent example of a circular oast with curved funicular walls stands at the Wye College Museum of Agriculture, at Wye, near Ashford, in Kent. By an incised date stone set high in the brick walls, the structure is known to have been built in 1815. Inside, one can understand at a glance how four brick hearths were fed from below a funnel that distributed the heat to the bed of hops above. Windows in the walls of the funnel at waist height allowed sulphur to be introduced in pans on the hearth.

In addition to encyclopedias and general texts on agriculture and gardening, a few imported English books specifically dealing with hop culture contained directions for building hop kilns. Edward J. Lance wrote one of the most comprehensive texts, entitled *The Hop Farmer; or, A Complete Account of Hop Culture*, first published in 1838, "to place the culture on scientific and rational principles."[16] The author noted that old kilns were long buildings about eighteen feet wide, with as many fireplaces as were required, each to handle two or three acres of the crop. The drying floors were ten or twelve feet square, grouped so that each would have access to the cooling room. "The hopper kilns when designed properly are constructed like two Pyramids placed base to base," Lance wrote. "Near the apex of the under one is placed the fire, and on the apex of the other is placed the cowl for ventilation." The drying floor was to be placed a foot below the union of the two bases.[17] Here the arrangement being put forward was not particularly new, but Lance's description was clearer than earlier ones. The fuel to be used was coke, charcoal, or coal, the last being the least satisfactory of the three because it often discolored the hops

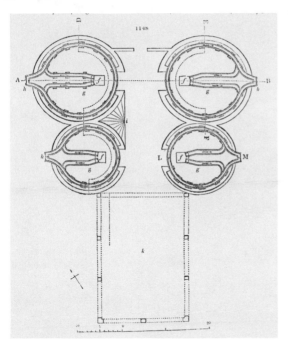

5.5 *Multiple oast scheme, Teston, Kent, England, 1832. Designed by John Read, the kiln builder to whom the introduction of the circular plan can be attributed, this multiple oast arrangement was one of the most sophisticated ever proposed up to that time. The plan* (left) *shows how four oasts were served by a common stowage room, while the section* (below) *demonstrates how the oasts functioned. (From J. C. Loudon,* Encyclopedia of Cottage, Farm, and Villa Architecture and Furniture *[London: Longman, Rees, Orme, Brown, Green & Longman, 1833], 597.)*

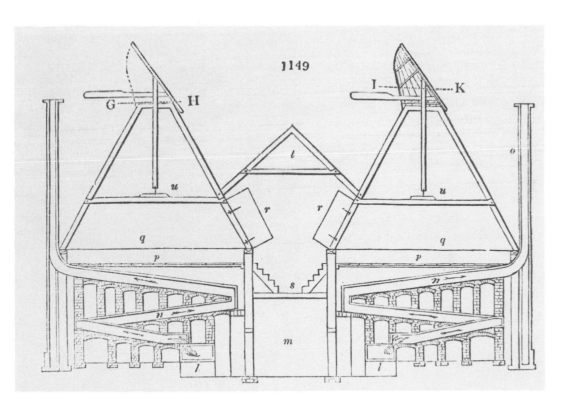

5.6 *Circular oast and stowage barn, Wye College Museum of Agriculture, Wye, near Ashford, in Kent, 1815. Here, on the interior of the oast (*right*), it is possible to see the curved funicular walls with one of the four ovenlike hearths.*

and created as much moisture in combustion as it drove off. Lance further noted that newer kilns employed furnaces that could burn any material because the hot air generated around the fire rose through the hops, while the smoke of the combustion chamber exited through a chimney.

A number of other types were employed. Growers often built simpler brick furnaces, three walls rising about a foot from the ground, on which rested an iron grate topped by a brick curb of six inches. The fourth side remained open for the removal of cinders. Other alternatives included a fire enclosure of iron plates with legs; experiments with duct work to regulate the intake and exhaust of air; the insertion of a double drying floor, the second of which would take the damp hops directly from the field; and schemes for drying hops with hot water or steam radiators.[18] It seemed that hop growers in Great Britain were actively exploring every avenue of invention.

Lance also commented on the rising popularity of round kilns, sixteen to twenty feet in diameter.[19] The construction of these "roundels," as they were sometimes termed, may have been encouraged by the assumption that the heat of the furnace chamber would be more evenly distributed over a circular floor. This belief was widely held at the time but proved to be incorrect, because the hops were being heated by convection—the movement of warm air through the space—not by radiation of heat from a point source. Nevertheless, the circular kiln with its high conical roof was one of the most distinctive building forms in the landscape. This remains true today, especially in the hop-growing regions of Kent, where they were often built in clusters.

Although English oast houses were sophisticated and information about their construction was available in the colonies, the earliest hop kilns on this side of the North Atlantic were comparatively crude. Perhaps the earliest published description is that provided by Israel Thorndike of Beverly, Massachusetts, who answered a plea for more information from the editor of the *American Farmer*, in Baltimore, in 1823. The details of construction which Thorndike provided are extremely revealing.

> A kiln for drying hops should be at the side of a hill or on rising ground, so that the top should be about nine feet from the bottom, twelve feet square at the top, tapering on all sides to about

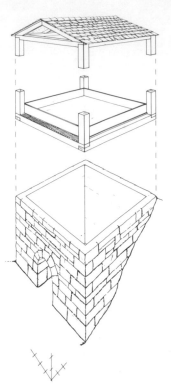

5.7 *Israel Thorndike's hop kiln, Beverly, Massachusetts, 1823. This exploded isometric, based on a meticulous description, demonstrates the parts of the earliest common kilns. The scale is marked in one-foot intervals.*

three and a half feet at the bottom *in the clear*, built up tapering, with stone laid in lime mortar, and plastered with clay from top to bottom with an aperture at the bottom about the size of the mouth of a common oven, for the convenience of putting in the coal, firing it, and regulating it afterwards.

Upon the stones on the top, is placed a sill of four pieces of timber of about eight inches square, and of course about twelve feet long, that being the size of the kiln at the top, upon which you place strips of boards, half inch thick and two inches wide, and within three and a half to four inches of each other, over which you stretch tow or coarse linen cloth, for a bed to place the hops upon, for the purpose of drying, and under which, at the bottom of the kiln, is made a charcoal fire, regulated at the discretion of the man who attends the drying. It will of course be necessary to have a board around the kiln at the top, of about one foot high, to confine the hops in the bed. I think it would be a further improvement to have a covered roof, and open at the sides, to protect the hops in case of rain, while they are drying.[20]

At the outset, then, a hop kiln is a vertical stone structure banked into the side of a hill, constructed in the manner of many lime kilns, for the convenience of those who are loading the hops. The mouth of the kiln provides the only draft inlet, and the fire is sustained by charcoal. The tentative nature of Thorndike's "further improvement" suggests that the idea of providing a roof was not frequently followed.

Was this scheme the exception or the rule in the hop-growing districts of Massachusetts? The evidence points to the latter. Later the same year, William Blanchard, Jr., a prominent grower in Wilmington, Massachusetts, provided a very similar description in the *New England Farmer*. Once again the side of the hill was specified as the best site, and stone walls constructed so as to provide a funnel twelve feet square at the top and two feet square at the bottom, eight feet deep. On top of the walls the builder laid the sills, joists, and a drying floor of lath, over which he stretched a thin linen cloth, nailing it in place. Boards about twelve inches wide, placed on the inner edge of the sill, formed a bin to receive the hops. Blanchard went on to describe the second story in greater detail than Thorndike, noting that "it is now common for those who have entered considerably into the cultivation of hops, to build houses over their kilns, which, in wet weather, are very convenient." Further, he stated that "if a ventilator was put in the roof of the building, directly over the center of the kiln, about six feet square, built like those in breweries and distilleries, I am of the opinion they would be found very advantageous."[21]

At first reading, the reference to building a house over a kiln may be seem unclear, but the early-nineteenth-century rural domestic structure was a straightforward building. Four sills, four corner posts, and four plates were the principal members, all on a stone foundation, topped by a gable roof. Blanchard goes on to mention that in cases where the "houses over the kilns are built large," for the purpose of storing the hops as they are dried, an air-tight partition should be constructed between the upper story of the kiln and this adjacent room, to prevent steam from injuring the dried crop.[22] Exactly how the kiln was to be attached to this room is left unspecified, although it may be surmised that they shared a common roof. If this was the case, perhaps the storage room was built on "stilts," that is, it extended from the front of the kiln and was supported at the other end by two posts. The importance of this reference, however, is that it demonstrates that the term *hop house* was only beginning to be used at this time, and that it referred to both the kiln and the cooling room.

Many of the earliest hop growers in New York State, men like James Cooledge and Solomon Root, were natives of Massachusetts and built in the manner with which they had become familiar. Little guidance was contained in published sources. For example, in 1801 the New York State Committee of the Society for the Promotion of Agriculture, Arts and Manufactures encouraged farmers to grow hops but could offer little help about the nature of the drying device except to note that a fire of charcoal should be used and the drying floor should be covered with a hair cloth in the manner of a malt kiln.[23] The fact that the first issue of the *Cultivator*, one of the earliest agricultural newspapers in the state, repeated the committee's report unchanged in 1834 suggests not only that the work of the society was highly valued but also that few advances had been made in the intervening decades.[24]

By the early 1830s interest in the problems of kiln construction increased so much that publications in central New York began to gather more information. The *Genesee Farmer*,

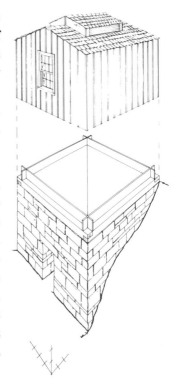

5.8 *William Blanchard's hop kiln, Wilmington, Massachusetts, 1823. This exploded isometric, drawn at the same scale as figure 5.7, shows how the second story might be enclosed. The scale is marked in one-foot intervals.*

published in Rochester, New York, sought ideas about the arrangement of kilns in Loudon's *Encyclopedia of Cottage, Farm, and Villa Architecture* and in the *Northern Farmer*, an agricultural paper published in Newport, New Hampshire. The *Genesee Farmer* also reissued Blanchard's 1823 article in the *New England Farmer*.[25] It should be noted that, as late as 1847, published correspondence confirms that the banked hop kiln was widely accepted as the proper form.[26] In addition, the scattered subsequent references that record the physical appearance of early hop kilns in central New York confirm that stone foundation walls, a slatted drying floor covered with a cloth, and a charcoal fire were commonplace components.[27]

As the number of growers increased, they tried various improvements and shared more specific information. The well-known grower Ezra Leland of Morrisville, Madison County, described kilns in his area in 1845. He noted that they were usually about eighteen feet wide and thirty feet long, and their first story was often brick or stone, about nine feet from the bottom to the first floor, with eighteen-inch-square holes in the bottom of the walls to let in air. Further, in such a kiln eight charcoal fires made at the bottom, equidistant from one other, provided the heat.[28] The dimensions of the kilns in this region seem to have been considerably larger than those to the east, for these examples had a drying floor that was about twice the area of the previous examples. They were also slightly higher, probably to avoid scorching the hops. More important, however, Leland made no mention of a funnel-shaped interior, nor did he say that the kilns were banked into a hillside. The introduction of ventilation holes on four sides indicates that the traditional, massive amount of stonework thought necessary to retain the heat was left behind. Instead the new generation of kilns had thinner walls, and the drying process relied almost completely upon the draft created by heating the incoming air to rise through the hops and escape above.

Exactly when growers adopted the pyramidal roof form to assist the draft is not known. One source indicates it was adopted near Waterville, New York, about 1850,[29] and this is generally consistent with what is known about its prevalence in the upstate hop-growing areas.

An example of a pyramidally roofed, banked, thin-wall stone

TINGED WITH GOLD

hop kiln stands on the west side of Valley Mills Road, a few hundred feet south of Haslauer Road, near Munnsville, Madison County, New York. In this instance a modest ramp leads up to the kiln, which has single square vents in its east and west walls at a level just below the drying floor, betraying an imperfect understanding of draft principles. A frame storage barn, demolished in the early 1980s, was connected to the kiln on the south, the whole arrangement reminiscent of that described by Blanchard.[30]

The thin-walled kiln, vented at the bottom of the walls, became complete with the introduction of the stove. As noted earlier, the use of a hop stove was not new to British growers, but not until the mid-1840s did blacksmiths in central New York go beyond providing plows and stoves for cooking and

*5.9  Hop kiln, near Munnsville, Madison County, New York, c. 1850. This is one of the earliest kilns in the area.*

parlor heating, to produce hop stoves and the necessary pipe for radiating heat beneath the drying floor.[31]

These developments—the thin-walled kiln, the pyramidal roof, and the hop stove—did not originate in New York. Time and time again growers looked farther to the east, beyond New England to Great Britain, for ideas that could be modified to fit local conditions. Descriptions of oast houses, circular in plan with masonry walls, appeared in the American agricultural press in the mid-nineteenth century, indicating the next step toward improving kiln arrangement. In 1847, for example, a traveling correspondent reported to the *Cultivator* on the circular kilns in Kent. These were from fifteen to eighteen feet in diameter, their conical roofs capped with swinging cowls. "The improved modern method," it was noted, had one or more large openings and fires in each kiln, "the draught being regulated by means of flues and sliding doors."[32]

Although New York growers built more circular-plan, conically roofed draft kilns than their brethren in other states, few survive. The example that stood on the James D. Cooledge farm near Bouckville was one of the most visible until the late 1970s, when it was razed. Unfortunately, because Cooledge was a very early grower, and this oast was built on the Cooledge farm, many authors have jumped to the conclusion that English forms were the earliest to be introduced in the region. As has been demonstrated, this was not the case.[33]

A larger extant oast house bears close examination. This structure, presently owned by Aloise Wroble, is located on the west side of State Route 8, about three miles north of Bridgewater, in Oneida County. The kiln is about thirty feet in diameter, and is linked by a bridge to a rectangular barn, twenty-five feet wide and thirty feet long. The walls of the kiln are cobblestone on the exterior and plaster over rubble on the interior. The barn is a braced frame structure with vertical board siding; the fenestration is much altered. Judging from the use of cobblestone and the form of the kiln and barn, the date of construction for the Wroble oast house is between 1845 and 1865. Two tension rings, one at the top and the other around the middle, are of uncertain date.

Although a conically roofed circular kiln is used in the same manner as the common type, the space between the kiln and the barn provides a key set of access points. The door to the first floor of the kiln, or stove room, is located beneath the

TINGED WITH GOLD

*5.10　James D. Cooledge's oast house, with later frame kiln, near Bouckville, New York, c. 1850(?). The cobblestone kiln with attendant frame barn was mistakenly believed to be the earliest kiln in the region. (From Marion Nicholl Rawson,* Of Earth Earthy *[New York: E. P. Dutton & Company, 1937], 129.)*

*5.11　Aloise Wroble's oast house, near Bridgewater, Oneida County, New York, c. 1850. The influence of English drying techniques is evident in this cobblestone building.*

second-story connecting bridge. There is also a doorway directly opposite leading to the first story of the barn, which may have been used in part for fuel storage. The hops, loaded through a door on the back (west) side of the bridge, could be cured in the usual manner, shoveled through the bridge into the top floor of the adjacent barn to cool, then funneled into the baling room below, emerging as a baled product at the double door on the south end of the structure.

Other oast houses can be found in the region, some with their cowls intact.[34] Despite the novelty and supposed advantages of the circular hop kiln, however, growers preferred the square form for its ease of construction.

The hop-drying building of the future would not be partly stone and partly wooden but rather entirely of frame construction. As early as 1847 Lincoln Cummings, a grower in Augusta, Oneida County, wrote that when building a hop house "the approved method would be to build it of wood, above the ground, being some ten feet below the eaves—the inner side of the walls lathed and plastered from the ground to the eaves," with a wood-fired stove supplying the heat.[35] The two chief reasons for this transformation were that a wooden building was less expensive to construct, and a frame structure already housed the supporting functions as part of the hop house. After isolating the fire in a stove, the most important function of the kiln was controlling the draft. This left little reason not to build the entire structure of wood.

By 1853, as described by one Otsego County hop grower, nine-tenths of the hop houses in his vicinity were four-room frame structures. The upper stories were devoted to drying and storing; the lower stories were used for a kiln and press room. To avoid scorching, he suggested, "In no case should the carpet on which the hops are spread be less than ten feet from the ground below."[36] The carpet could be of cotton or linen, well stretched and firmly tacked to the rack. The kiln, being sufficiently high to stand in, was to be plastered to the plates to prevent outside winds from entering. Stoves were quickly replacing furnaces and open fires as the best way of heating the kiln. In all, while a few, established growers continued to favor the use of a banked stone kiln,[37] by the mid-1850s progressive farmers had adopted the frame hop house as the most modern and cost-effective structure for the purpose.

The shift from a masonry kiln to a frame kiln produced hybrid

TINGED WITH GOLD

buildings, at least one of which survives. The hop house on the property of George W. Reilly, located on the north side of Tower Road, about three-quarters of a mile west of Waterville in Oneida County, New York, is important as a rare example of a braced frame building with brick nogging in its kiln. The structure stands on a level site, about fifty feet from the nearest barns, which screen it from the temple-fronted Greek Revival–style house, a comfortable distance that abated the potential fire hazard and the noxious smell of brimstone or sulphur used in the curing process. The pyramidal-roofed draft kiln is nearly square, measuring about twenty-two feet long and twenty-one feet, ten inches wide. Access from the outside is provided on the first-story level by a door to the furnace room on the front (south) facade and from the interior by a door to the pressing room. On the second-story level, a break in the cornice line on the west facade locates the former hop loading door, while on the interior the doorway to the cooling room, opposite, remains. Once inside the kiln, one can observe that the walls and the ceiling were completely finished in plaster on sawn lath. The brick nogging fits snugly between the studs on all four walls, retaining the heat. The three-bay cooling room and pressing room are also notable, for one of the hop chutes in the floor, through which the hops were shoveled to the press below, still remains.

With the image of the relatively intact Reilly hop house firmly in mind, an example in Otsego County is understandable. Only the pyramidally roofed kiln remains of this hop house, standing on the John Moakler farm on the east side of County Route 33, about three miles north of Cooperstown. The barn, a three-bay-wide, two-story frame structure attached to the north, was dismantled in the summer of 1976, leaving the kiln a freestanding building. From examination of the kiln alone, however, the juxtaposition of the four principal rooms in a typical hop house of the period is very clear.

Whereas most of the traditional methods of construction depended upon a frame of heavy square timbers with mortise-and-tenon joints, these methods were superseded by one of the most significant developments in American building technology, the "balloon frame." The distinguishing characteristics of this framing technique were that it relied largely on two-by-four-inch studs, joined by nails, extending from the foundation sills to the eave plates to carry the load from above. This

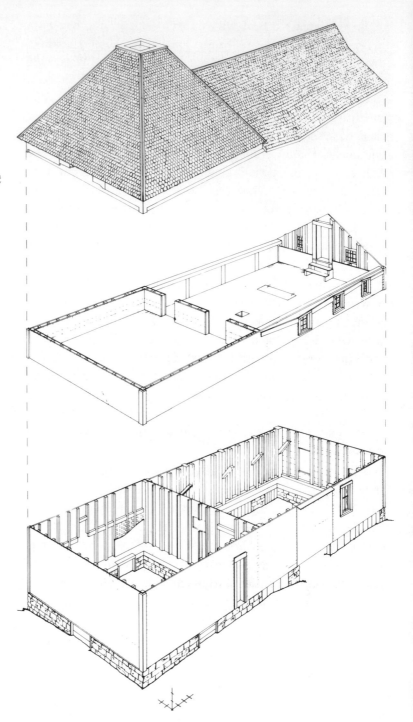

*5.12  George W. Reilly's hop house, Tower Road, near Waterville, Oneida County, New York, c. 1850(?). This exploded perspective shows the juxtaposition of the rooms, and the braced frame construction. The scale is marked in one-foot intervals. (Measured drawing by Carlos Villamil-Herrans, with assistance from Suzanna Barucco and Stephen Callcott, 1987–88.)*

*5.13  Reilly hop house, nogging detail. The brick nogging in the kiln walls was plastered over.*

"basketwork" of sticks, although at first the subject of derision, was far easier to construct and proved stronger than was first believed. After being introduced in Chicago in 1832,[38] it was widely adopted throughout the Middle and Far West, and gradually became accepted in the East. The introduction of the balloon frame allowed inexpensive frame buildings of any type to be erected almost anywhere that the idea was understood and the materials could be shipped. As Otsego County grower J. W. Clarke wrote in 1860:

> Hop drying pre-supposes a drying kiln. In their best form kilns are costly. Having used and seen others use a cheap kiln, I will here briefly outline its form for the uninitiated, if such there be, who feel any interest in the subject. This can be available only to those who have a large barn or other floor on which to lay the heaps, as they are dried: Build a wall three feet high, 16 feet by 12. In the center of 12 feet square, in one end of this basement, build a solid block of masonry 6 by 6 feet, and 18 inches high. Now raise a light balloon frame, 16 by 12 feet high, on the basement wall; and put on this a roof of one-third pitch or grade. A small frame of 2 by 4 oak scantling is now laid on the central

TINGED WITH GOLD

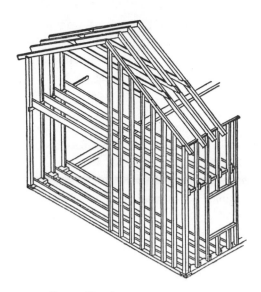

5.15   *George Woodward's balloon frame, 1860. Woodward demonstrated to the agricultural community in one of the foremost agricultural journals of the day the viability of this midwestern framing technique. (From "Balloon Frame—III,"* Country Gentleman *15, no. 14 [April 1860]: 229.)*

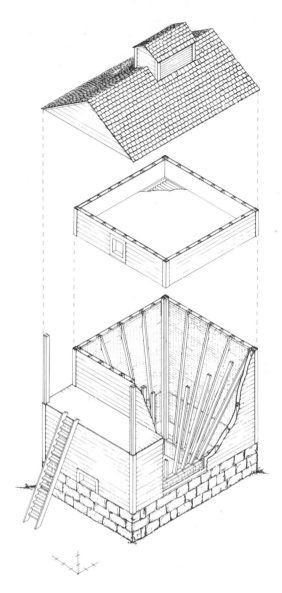

5.16   *J. W. Clarke's balloon frame kiln, Wisconsin, 1860. This perspective shows the manner in which Clarke proposed constructing this kiln, independent of an attendant barn. The scale is marked in one-foot intervals.*

block, having an inside space of four feet by four. Oak studding are laid on this, and carried up—spreading, funnel-like—till they reach a perpendicular height of 8 or 9 feet in the clear. Here oak joists, 2 by 6, cross from the studding to the studding of the frame. On top of these are laid strips of pine, or other wood that is not apt to warp so much, one and one-half by 3 inches the deepest downward, with one and one-half inch spaces between them, making a "spaced" drying floor 11.8 by 12 feet. The space of four feet on one side is floored five feet above the wall, and four feet below the top of the inclosing frame of the kiln. This floor is necessary to stand on in examining the hops when drying, taking them off, or putting them on, & c., and is accessible from the outside by a step ladder. A ventilator in the center of the roof ridge is indispensible. On the side in the basement where the four feet space is left, an opening 15 by 18 inches is left in the frame, the bottom of which is the block. The frame being lined with inch board previously, the opening and interior is lined with brick, set in well-tempered mortar; the brick being laid the thickest ways for the first three feet, and then flat against the boards to the top. Now the kiln is complete; the space around the block in the basement serving to store charcoal in, and the object of the wall being that of security against fire. Two kilns are necessary to dry hops economically, and not more than one acre—good hops—must be attempted with only one kiln. The cost of such a building is, I think, about $40 in Wisconsin; and it is useful for many other purposes than drying hops, though I have not had time to specify.[39]

Several points need to be underlined. First, Clarke's initial observation—that such a kiln can be used only in tandem with a large barn or other floor on which to lay the hops after they are dried—explains, in part, why this model is so inexpensive. As has been shown, early kilns were independent of cooling barns. The balloon frame made construction easier. Second, the vertical dimensions of the kiln, from the floor to the bottom of the drying floor, have increased to eleven or twelve feet. Although the balloon frame would have allowed for a greater height, this increase is significant for guarding against scorching the crop. Later balloon-frame kilns were even taller. Third, while the use of charcoal for heating in a funnel-shaped interior is clearly a *retardetaire* approach, it may also be an attempt to avoid the expense of a hop stove and piping.

An examination of hop houses in central New York reveals that as hop growers rebuilt or expanded their operations, they incorporated the balloon-frame technique as a matter of course.

One such example is on the farm of Michael Slater, on the south side of Otsego County Route 52, about one mile east of the intersection with County Route 33. This is a two-part structure. The kiln, on the south, is apparent because of the air vents at the bottom of the fieldstone foundation. When viewed from the road, the differentiation is clear between the horizontal clapboards over a balloon-frame structure and the vertical planks on the north end of the barn. The nine-over-six double-hung sash window in the south gable was used in lieu of a ventilator or cowl, while two more double-hung windows on the east and west facades admit light to the pressing and baling room. On the east (rear) facade, near the eaves, is a loading door by which hops were admitted to the drying floor, making use of the hill, while on the north facade a large barn door allowed the bales to be loaded onto wagons that would roll down the drive.

*5.17 Michael Slater's hop house, Route 52, near Cooperstown, Otsego County, New York, c. 1850 and c. 1895. An old kiln was often expanded with a new barn or hop house to increase the drying capacity.*

HOP KILNS, HOUSES, AND DRIERS

As a whole, the Slater hop house might be termed a remarkably intact, straightforward example of a common hop house. However, a closer analysis reveals even more about the building. The balloon frame of the kiln does not match the earlier, braced framing of the barn. Further, cut-wire nails were used in the construction of the kiln, dating it after about 1895. The barn contains evidence of an earlier kiln located in the middle of the present structure. There are indications that studs once partitioned the interior of the barn. Evidence of the old kiln is further confirmed by plaster on split lath nailed with flat, sheet-cut nails on the east and west walls, and by the apparent reconstruction of the foundation. Thus, the simple form of the common hop house is not only the earliest type but also one of the most long-lived solutions.[40]

The relationship between the kiln and the barn in the late-nineteenth-century hop houses of New York State always remained a close one. However the juxtaposition of these two elements might vary considerably. The hop house of Robert Snitchler, on the east side of Smith Road near the village of Hamilton, in Madison County, provides a good example of how the plan could change. Originally, the Snitchler hop house consisted of a three-bay-long barn with an attached draft kiln on axis, conveniently facing the fields. Today, the building that appears as only the barn was once both the kiln and the barn, the most obvious external clues being the break in the form of the building, and the use of horizontal board siding on the kiln to differentiate it from the vertical siding of the barn. Although the distinctive pyramidal roof of this kiln has been removed and its second story rebuilt to accommodate a new metal roof, the confirming evidence includes the vents in the foundation and the plaster-over-lath finish on the interior.

Sometime after the original hop house was constructed, a second draft kiln was added to the north, next to the east facade of the barn. This may have been done either to increase the total drying capacity of the hop house or to allow the barn to be enlarged by incorporating the old kiln. In either case, the kiln was a near duplicate of the old one. This suggests that the size of the kiln was determined by its drying capacity, a conclusion that can be reinforced by looking at other examples.

A number of agricultural periodicals followed the growth of the industry, as an increasing number of farmers requested information. A few journals included a great deal about hop-

5.18 *Robert Snitchler's hop house, Smith Road, near Hamilton, Madison County, New York, c. 1880 and c. 1895. This hop house was expanded by the addition of a draft kiln to one side.*

house design and provided models to follow, largely because of the particular interest of the editorial staff. In November 1864, for example, the *American Agriculturalist* offered a prize of seventy-five dollars to "practical men" who could convey the greatest amount of information about hop culture in the smallest amount of space. The illustrated essays which were submitted and found useful were collated into a book entitled *Hop Culture*, one of the best sources of the period. Andrew Samuel Fuller, the editor, wrote a brief introduction in which he noted that each of the nine essays demonstrated some details of practice different from the others.[41]

The prize-winning article, by Heman C. Collins of Morris, Otsego County, presented perspective plans of two hop houses that had identical floor plans. The first was termed the "common" variety. A four-room arrangement was shown: the stove room, with stone, brick, or plastered walls, but no floor; the drying room with plastered walls and a floor composed of one-by two-inch slats; the storeroom, which had a window in the end with tight shutters; and the baling room, which was generally well lit. Other distinctive features included the ventilator over the drying room; the platform from which the green hops were loaded into the drying room through a large door; the stovepipe or smoke stack, which was to be taken down when not in use; and the air-intake holes, each measuring one by three feet, with shutters to close them when necessary. Particularly noteworthy are the dimensions given across the exte-

5.19  Heman C. Collins's hop house, Morris, Otsego County, New York, 1864. A, *stove room;* B, *drying room;* C, *storeroom, window not shown;* D, *baling room;* E, *ventilator;* F, *loading platform;* G, *door;* H, *stovepipe;* I, *air-intake holes;* a, *stove;* b, *heating duct;* c, *drum;* f, *floor-level vents;* x, *trap door. (From E[zra] Meeker,* Hop Culture *[Puyallup, Washington Territory, 1883], 7.)*

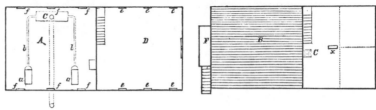

rior of the building, indicating the division of the hop kiln from the hop barn, and likewise showing the floor level of the four rooms.

The ground plan illustrates two stoves with long heating ducts stretching across the stove room to a sheet-iron drum. The flue rose from this drum and returned across the room to provide for maximum heating before it exited. The stack was to be kept away from any woodwork, so as not to set it afire. There were six vents at floor level. The adjacent baling room contains only the stairs to the storeroom, above. Three windows lit each side and a double door in the gable end permitted easier unloading of the baled product.

The second floor plan indicated very clearly the slats of the drying floor over which a loose-knit carpet would be placed. The slats were perpendicular to the joists but, more important, were parallel to the direction in which the hops, once cured, would be shoveled into the storeroom. After the hops cooled they would be passed as needed through a trap door to the room below, for baling. The illustrations and the accompanying description of the structure were very helpful to the interested reader.

The second perspective view presented in the Collins essay was labeled a "draft" variety. In this instance a tall pyramidal roof with cowl was substituted for the gable roof and ventilator. The cowl was mounted so it could rotate, directed by the wind pressure upon a vane. The plans of the building, however, were unchanged.[42]

"E.O.L." of Vernon, Vermont, illustrated another hop-house arrangement in the second prize essay. In this case the builder banked the structure in the traditional New England manner so that the construction of a ladder and platform to deliver the hops to the drying room was unnecessary. This hop house was twenty-two feet wide by thirty-two feet long, with a kiln sixteen feet square and a walk around it. As in the previous schemes, the drying floor was built higher than the floor of the storeroom.[43]

Still another example of a hop house was put forward by A. F. Powley of Squankum, Monmouth County, New Jersey. Following the English practice, the author referred to his kiln as

5.20 Collins's draft hop house, 1864. The tall pyramidal roof gave the hop house an easily recognizable form. (From E[zra] Meeker, Hop Culture [Puyallup, Washington Territory, 1883], 10.)

HOP KILNS, HOUSES, AND DRIERS

185

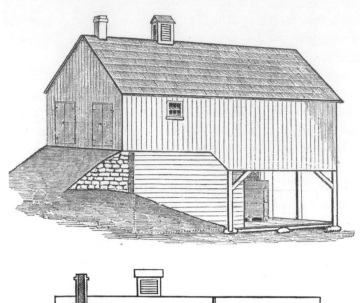

*5.21 E.O.L.'s banked hop house, Vernon, Windham County, Vermont, 1864. The traditional New England arrangement used a natural slope to assist in providing access to the second story of the kiln. (From E[zra] Meeker,* Hop Culture *[Puyallup, Washington Territory, 1883], 14.)*

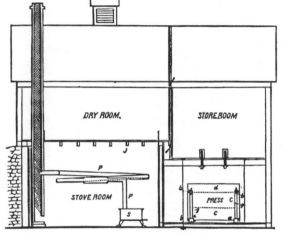

an "oast," and he proposed a circular kiln, about twelve feet high. As before, a stove provided heat, and several intake holes were cut in the wall near the floor of the kiln. Joists placed in the wall at a height of about eleven feet carried slats covered with a strong cloth stretched to carry the hops. Above the drying floor the wall was raised a few more brick courses, and a tall, conical roof was erected.[44] The first floor of the adjacent barn contained coal bins and the press.

The last example to be included in *Hop Culture* was a double oast proposed by Albert W. Morse of Eaton, Madison County, New York. This design was to serve a large yard. The kilns were round in form, and could be built in brick, stone, or wood. If the builder chose wood, Morse proposed a balloon frame as the most convenient. Like the examples proposed by Heman

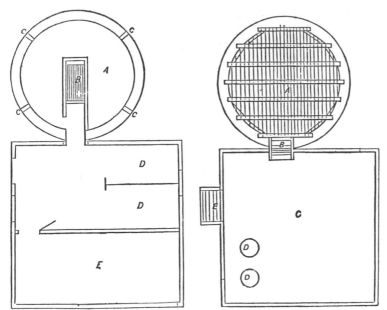

5.22 *A. F. Powley's oast house, Squankum, Monmouth County, New Jersey, 1864.* Left: A, *stove room;* B, *stove;* C, *draft hole;* D, *coal bin;* E, *pressing room.* Right: A, *floor covered with slats;* B, *steps to storeroom;* C, *storeroom;* D, *hole in the floor for baling. (From E[zra] Meeker,* Hop Culture *[Puyallup, Washington Territory, 1883], 21.)*

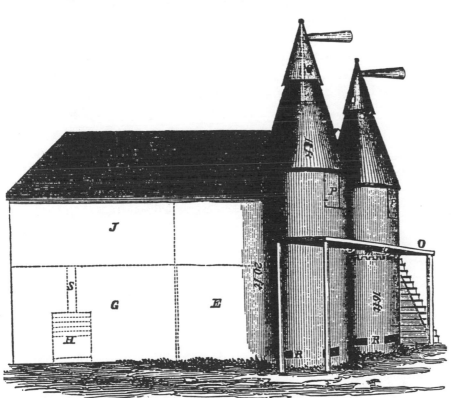

5.23 *Albert W. Morse's double oast house, Eaton, Madison County, New York, 1864. (From E[zra] Meeker,* Hop Culture *[Puyallup, Washington Territory, 1883], 25.)*

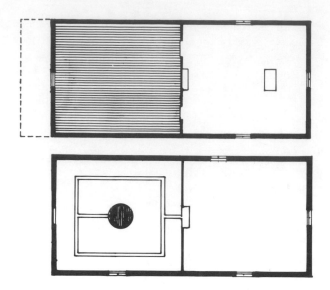

*5.24  H. H. Potter's hop house, Sauk County, Wisconsin, 1867. The press and storeroom (bottom) requires a tight floor. A door leads from this room into the kiln. The upper storeroom (top) is reached by a flight of stairs from the press room. A slide or trap door in the floor allows the hops to pass down into the press room. To increase the draft, this type of balloon frame structure could be built with either a gable or high pyramidal roof. (From H. H. Potter, "Hop Boxes and Hop Houses," Prairie Farmer, n.s. 19, no. 12 [March 1867]: 183.)*

Collins, a platform at the top of a flight of stairs was needed to charge the kilns.[45]

The buildings illustrated and discussed in these four articles indicated that a wide range of forms and arrangements had developed. In addition to being a mirror of this progress, *Hop Culture* was also undoubtedly an important guide for potential growers.

By the late 1860s midwestern periodicals, such as the *Prairie Farmer*, also began to furnish their readers with specific information about the erection of hop houses. For example, in March 1867, H. H. Potter of Sauk County, Wisconsin, provided the specific details of his four-room, balloon-frame hop house, constructed during the summer of 1865. It was a freestanding structure, measuring twenty by forty feet, with studding sixteen feet high, "under one roof, quarter pitch, at a cost of $400." Potter also noted that "in another common style of hop house the kiln is higher between joists, with a four-square roof running to a point, surmounted by a ventilator, with press and store room forming a wing with common roof."[46] The description and the accompanying illustrations were so important that they were incorporated in the equivalent of the *Hop Culture* text published for Wisconsin growers in 1868, a booklet entitled *The Cultivation of Hops, and Their Preparation for Market as Practiced in Sauk County, Wisconsin,* by D. B. Rudd and E. O. Rudd.[47]

A close look at Potter's letter and other articles in periodicals of the time indicates that Wisconsin growers were well aware of all the latest technological inventions employed in the eastern United States and England. What becomes so striking about the hop craze in the Badger State is that it was such a fertile period of experimentation.

It should be recalled that the pioneer hop grower of Waukesha County was James Weaver, who moved from New York State to Wisconsin in 1837. Unfortunately, no description of his kiln survives. In Sauk County, Jesse Cottington, an English immigrant who had worked in New York State, planted his yards in 1852 and built his log kiln shortly thereafter. Its measurements were twelve by twenty-four feet, and charcoal was burned in the drying process.[48] Such rudimentary kilns may well have been more the rule than the exception during the 1850s and early 1860s, but improvements were made quickly. The undated Canfield kiln in Baraboo, an octagonal structure of

5.25  *Canfield hop kiln, Sauk County, Wisconsin, c. 1868(?). A log structure of this kind obviously mimics the English oast form but is constructed of materials at hand. Photograph taken c. 1925. (Sauk County Historical Society, Baraboo, Wis.)*

logs, survived long enough to be included in the county history,
an obvious imitation of the more fashionable albeit more expensive stone- and rubble-wall oast houses, such as the one built
about 1854 in Wyocena.[49]

The variety of kilns in Wisconsin soon equaled the range of
devices in New York, in part because any technological improvement in production introduced in one hop-growing region
was almost immediately in demand in the others. For example,
W. A. Brooks of Oneida County, New York, a grower who had
been "extensively engaged in building hop houses, and cowls,"
in the hop-growing regions of that state, worked in the Kilbourn City, Wisconsin, area for a number of months in 1868.
Brooks was responsible for "putting up, in this region, the
English cowl," which had already been noted as a standard
feature in the East.[50]

Experimentation with materials led to unusual solutions. In this respect, the Beaumont hop kiln on Rybeck Road in Hartland, Waukesha County, Wisconsin, is a unique survivor. Reminiscent in form of the balloon-frame kiln described by J. W. Clarke, this structure was apparently built without an attached barn. Eighteen feet square at the base, it stands thirty-two feet high with a tall pyramidal roof. A doorway in the center of the east elevation gives access to the lower level; three openings were made at the bottom of the remaining walls for ventilation. The chimney rises in the west wall, suggesting that the heating source was a stove. Access to the upper story, the hop-drying room, must have required a ladder or a platform with a flight of steps that led to a smaller door over that below.[51] A chute in the grillwork of the drying floor allowed the cured hops to be dropped into sacks or a press.

The Beaumont kiln is not a frame building; rather, its walls

5.27  *Ephraim Beaumont's hop kiln, Hartland, Waukesha County, Wisconsin, c. 1863. Although this example is one of several typical forms in the region, it is more important for its early concrete walls.*

are of pisé construction. The structure is constructed of field-stone in a hydraulic cement mortar mix—a crude concrete—which was built up in successive layers using wood planks as formwork. The forms were reused and elevated until the walls achieved the desired height. Because Ephraim Beaumont built the hop kiln on his farm sometime after he purchased it in late 1862 or early 1863, and because the kiln appears on an 1875 county atlas, the date of construction lies firmly within the peak years of hop production in Wisconsin.[52]

Although most of the small frame kilns that characterized the midwestern landscape have since disappeared, sufficient physical evidence remains to determine their dimensions.[53] A few Wisconsin growers built large, well-ornamented hop houses, however, as is evident from the insurance settlements paid after the structures and their contents went up in smoke.[54] The twin hop houses of the Ruggles family, constructed in 1867 and razed in 1902, are characteristic of the period.

In the post–Civil War period the focus of attention was not a change in the cowl, the wall materials, or the form of the hop kiln. Rather, as the need to accelerate production arose, inventors began to address a more immediate problem: the heat that was wasted while men moved the cured hops off the drying floor and loaded the green hops.

In New York State, Edward France addressed this concern with his Improved Hop-Dryer, patented in 1864. In the France kiln the cloth drying floor could be moved by means of a lever

*5.28  Ruggles family hop houses, Sauk County, Wisconsin, 1867–1902. These twin frame structures suggest that some Wisconsin hop buildings were of considerable size. (Sauk County Historical Society.)*

TINGED WITH GOLD

5.29 *Edward France's "Improved Hop-Dryer," Cobleskill, Schoharie County, New York, 1864. Dried hops were removed by means of a movable cloth, thus saving fuel and time "without losing a particle of hop dust." (From E[zra] Meeker, Hop Culture [Puyallup, Washington Territory, 1883], 8.)*

applied at the end of a shaft or roller, on which the cloth was wound. Testimonials signed by many of France's Cobleskill neighbors claimed that the grower needed only four to six minutes to move the whole kiln of dried hops from the main carpet into the storing room and return the carpet to the original location ready for loading more green hops.[55] Further, the device diminished the amount of heat lost and lessened the discomfort of working around the heated kiln.

Whether France's invention ever became widely employed is not known. However, the race was on to perfect a system whereby the hops could be moved en masse from the drying room to the storeroom. Mechanization seemed to be the way of the future, and Wisconsin growers forged ahead with the idea that the hops could be carried through the dryer on movable trays. Jonathan Whitney of Port Hope, Columbia County, was one of the earliest to develop and build a "dumping hop floor." In 1868 Whitney patented a large shallow box that ran on a pair of tracks between two rooms inside a gabled two-story structure, which measured about seventy feet long, twenty feet high, and twenty feet wide.[56] Hops placed on a screen over a stove and stovepipes in the usual manner were, when dry, slid on the tray to the adjacent room and dumped to cool, while the

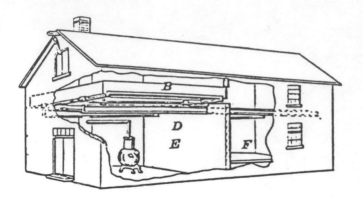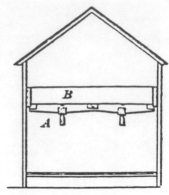

5.30 *Jonathan Whitney's hop drier, Port Hope, Columbia County, Wisconsin, 1868. Whitney's invention of the dumping hop floor went further than France's invention by providing a pair of tracks for a tray to move upon. A, drying floor; B, shallow box with sides; D, track, or guides; E, stove room; F, storeroom. (From* Annual Report of the Commissioner of Patents . . . 1868 *[Washington, D.C.: Government Printing Office, 1869], III, 325.)*

box could be returned to its place in less than a minute for reloading. This invention was significant because the mechanized drying process determined the form of the hop house. Further, the idea became a reality in a number of locations, for growers actually built Whitney's hop houses in Sauk and Columbia counties.[57]

The idea of the dumping hop floor soon spread. In 1869 William Loofbourow of Lafayette, Wisconsin, patented a system whereby pans containing the hops could be moved on a track between two rooms, the contents being emptied by means of hinged sides in each pan.[58] About 1871 Daniel Flint, the prominent California hop grower, constructed a kiln that incorporated a movable drying floor with two dumping areas. In this instance the drying floor was one-third the length of the whole building, and the other two-thirds were two storage bins, one on each side. The drying floor was a car in the form of a frame with a wire cloth bottom, moved to either side by means of a windlass. When the frame was raised up on edge, the hops slid off into the bin below. The advantage, Flint claimed, was that the car dumped the hops off in layers evenly over the cooling rooms and allowed each lot to cool before the next was put on top.[59]

All of these inventions, however, called for the process to be completely contained within the hop house. W. F. Waterhouse, of Weyauwega, Waupaca County, Wisconsin, became the first to propose a system that separated the kiln from the cooling room. He patented the scheme in 1868.[60] The hops, loaded at the door to the drying floor, would be placed on the drying floor inside the kiln until cured. They would then be carried on the movable floor over the tracks and pushed into the cooling room

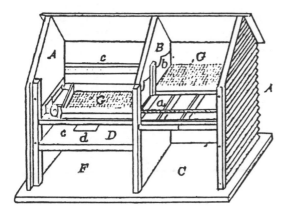

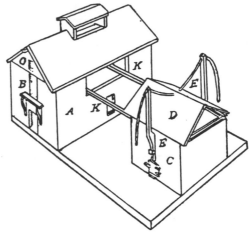

5.31  William Loofbourow's hop drier, [La]fayette, Monroe County, Wisconsin, 1869. a, stationary, slatted floor; b, gate lifted to allow pans to move; c, track on which pans move; d, hole to storeroom; A, exterior wall; B, interior wall; C, stove room; D, cooling room floor; F, storeroom; G, pans on which hops are placed. (From Annual Report of the Commissioner of Patents . . . 1869 [Washington, D.C.: Government Printing Office, 1871], III, 338.)

5.32  W. F. Waterhouse's hop drier and cooling barn, Weyauwega, Waupaca County, Wisconsin. The cured hops were carried over tracks to an adjacent barn. A, cooling rack; B, loading platform; C, kiln; D, kiln roof; E, ropes to lift roof; K, tracks; O, loading door. (From Annual Report of the Commissioner of Patents . . . 1868 [Washington, D.C.: Government Printing Office, 1869] IV, 982.)

under the roof, which would be opened by means of ropes. Although the operation of the roof on the cooling room is rather awkward, the principle of separating the two main functions of the hop house, drying and cooling, was a wise one. This safety feature reduced the possibility of fire, or at least allowed fires to be contained more easily. Growers soon learned that if preventative measures were taken, insurance rates might be reduced. This was no small consideration, particularly among commercial operators who wished to build larger structures.

In retrospect, perhaps the most important contributions of Wisconsin's comparatively brief hop-growing effort lie in the introduction of the balloon frame and in the preoccupation with mechanical drying. A little over a decade later both of these improvements were adopted by the large growers on the Pacific Coast.

The earliest kilns built on the West Coast were not very different from the contemporary structures in the East. Daniel Flint, recognized as the first grower in California to build a kiln, erected a frame building for the purpose before 1861. It contained on the ground floor a brick oven, from which heated air was conveyed through pipes beneath the slats of the drying floor, in the usual manner.[61] Six years later, William M. Haynie of Sacramento, in an address to the State Agricultural Society, described the hop houses commonly being built in his region as four-room frame structures "usually built under one roof," measuring about twenty feet wide by thirty-six feet long.[62] A number of hop houses were built by Chinese tenant farmers, but the slight descriptions that have survived do not indicate any improvements that might have been made over previous models.[63]

Perhaps the first mention of an innovation occurred in a model structure described in the *Pacific Rural Press* in 1874. This example, deemed suitable for drying the hops produced by a yard of about ten acres, was a building eighteen by thirty-six feet, with posts sixteen feet high, divided into four rooms.[64] In place of a wooden grid floor between the stove and drying rooms, a heavy wire cloth was fixed to the joists, and over this a baling cloth. The use of a wire screen instead of wooden scantling became a distinctive characteristic among large Pacific Coast growers, while the outside form of the hop house remained unchanged.

By the late 1870s production had apparently risen high enough on some plantations that growers built groups of large hop houses. For example, Thomas Clement of Poland, San Joaquin County, had three hop houses, the largest a two-story brick structure, thirty by ninety feet, sitting on the brow of a hill, and two others of frame construction some distance apart, thirty by fifty and eighteen by forty feet respectively.[65] It remained the practice on small farms to meet the demand for increased drying and storing capacity by building additional four-room frame structures well into the twentieth century. The tendency on large establishments differed. After the early 1890s, when a new wave of structures was built,[66] growers in the Sacramento Valley built their kilns completely separate from their cooling barns to minimize any damage by fire. Glacken and Wagner, Ezra Casselman, Mrs. N. A. Bendler, W. H. Leeman and Lovedale Brothers, Palm and Winters of Walnut Grove, Flint and Raymond and W. F. Warburton of Elk Grove, and George Menke, all constructed new hop kilns with separate cooling rooms. Farther south, the Pleasanton Hop Company grew to having twelve kilns, each with a thirty-by-thirty-foot drying floor, all connected with overhead trestles to two large cooling warehouses. The cars that shuttled between the kilns and the warehouses measured thirty by twelve feet, large enough to take an entire floorload of hops at once.[67]

The concern for fire prevention can be seen not only in the layout of the yards, where the kilns were separated by a considerable distance from the cooling barns, but also in the choice of construction materials and new techniques. About the turn of the century, at Horstville, on the banks of the Bear River in Yuba County, the Horst brothers constructed kilns and cooling barns entirely of corrugated sheet iron, with frames of "bridge construction." The latest improvement in their method was to dry hops at the lowest possible temperature. Fan blowers drove currents of air through the hops without artificial heat of any kind.[68] Soon the company remodeled all its other kilns to conform to this battery of six new kilns, but no evidence has come to light suggesting that the Horst technique was widely adopted by other growers.

Although other technological advances did gain widespread acceptance, the road to successful hop drying in the future was not always clear. Commercial growers who did not want to invest in separate kiln and cooling buildings, which required tracks or rails with trays or wagons, looked for other means of expediting the process of drying. In the early 1890s, for example, Patrick Cunningham, one of the largest growers in the Ukiah vicinity, began to substitute a coarse screen similar to fishing net for the burlap carpet that covered his wire drying floor. When the hops were dry, the net was taken up in sections and emptied on the floor of an adjacent cooling room. Cunningham's kilns were also unusual because they contained a second drying floor directly above the first, doubling the capacity with allegedly no increase in fuel and a great savings of time and labor. On the other hand, he also believed that hops should be dried slowly to better preserve their essential qualities and thus apparently avoided using the fan-blast process favored by others.[69]

The early hop-drying buildings in the Pacific Northwest appear to have been little different from contemporary structures elsewhere. The first hop building in Oregon about which something is known was a four-room frame structure built in Santiam Bottom, Marion County, Oregon, by William Wells by 1869. Wells's kiln was eighteen feet square, "with furnace" below.[70]

Only four years earlier, in the Puyallup Valley of Wash-

5.33 Patrick Cunning-
ham's kilns with barn,
near Ukiah, Mendocino
County, California, 1909.
These kilns were some-
what unusual for having
two drying floors each.
This is confirmed by the
two-story loading plat-
form. (Mendocino County
Historical Society, Ukiah,
Calif.)

ington, Ezra Meeker watched his father's first attempts to cure
a crop in the loft over the kitchen fire.[71] The elder Meeker's
first hop house was built of logs, "chinked and daubed," with "a
furnace to heat it, built of clay, flues of the same material
covered with sheet iron, and an old-fashioned 'cat and clay'
chimney."[72] Meeker also remembered his father's look of de-
spair when, in the course of attempting to make the "new hop
house work," he found the upper room filled with fog and
condensation dripping back on the hops. The problem was
corrected by opening the foundation vents and providing a
tighter roof, to increase the draft. Buildings on the frontier
were always adaptable.

Undoubtedly the Meekers learned something from Levi F.
Thompson, the most successful pioneer grower in the Puyallup
Valley.[73] In 1867 Thompson erected his first drying house, a
structure twenty feet wide and seventy-five feet long, which
included a kiln at each end.[74] Apparently he had little question
about how to build a kiln, and although evidence is scant,
Thompson may well have been the most knowledgeable builder
in the valley.

By the fall of 1874, seven of the eighteen growers in the seven-mile-long and two-mile-wide hop-growing area along the river bottom had built their own "drying houses or kilns." Those growers who did not take the trouble sold their crops standing in the field.[75] Recognizing the need for additional facilities, in 1877 Thompson erected five drying houses, each costing between a thousand and twelve hundred dollars, the largest single investment in the territory to that time. The last two kilns, it was reported, were painted and ornamented, and one, "if it had but the windows, would be mistaken by a majority of strangers as a country academy."[76]

About this time the buildings known as hop houses elsewhere in the country became known as "hop driers" or "dry houses" in Washington Territory. The terms reflect no difference in the construction or the curing process.[77] The gradual loss of the term *kiln*, however, indicates the evolution of the building type, divorced completely from the tradition that had spawned it. Henceforth, the emphasis was always on providing the proper kind of convection. Growers farther south, in Oregon, subsequently referred to their structures in the same fashion.

*5.34   Early kiln, Puyallup Valley, Washington Territory, c. 1888. Such structures required little cash outlay; only the time and labor of the pioneer was necessary to construct them. (From "Hop Pickin' in Puyallup,"* Harper's Weekly *32, no. 1661 [October 1888]: 801.)*

A good example of a prosperous hop ranch in Washington was Darius M. Ross's 343-acre farm about one and a half miles west of Puyallup.[78] In 1875 Ross erected his first drying house, a structure sixty feet square, with two kilns side by side, each eighteen feet in diameter. The subsequent enlargement of his fields engendered a storage addition. A second drying house was built in 1884. It measured eighty-two feet long by twenty-eight feet wide, with a taller pyramidal-roofed kiln on one end, a receiving floor twenty-eight feet wide and fourteen feet long in the center, and a storehouse and press room on the other end.[79]

Much more ornate than Ross's buildings was the drying house of J. R. Dickinson of Sumner. In this instance the twenty-four-foot-wide and ninety-six-foot-long building, which had a forty-foot-long storage and press room to the north, was accented with an octagonal cupola on each ventilator, capped with a decorative spire.[80]

The visual differences between kilns were often as much an indication of technological innovation as a reflection of stylistic choice. On the Ross ranch, for example, what could account for the change from a tall, thin ventilator on the 1875 kiln to a shorter version on the 1884 drier? The answer can be deduced from an 1882 cut of Ezra Meeker's kilns, which shows three older draft kilns with tall ventilators built alongside his new

TINGED WITH GOLD

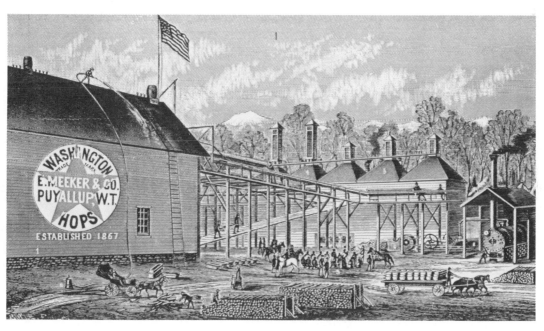

5.37  Ezra Meeker's operations, Puyallup,
Pierce County, Washington Territory, 1878 and
1882. Three old draft kilns with tall ventilators
stand alongside the new fan-blast kilns. (From
E[zra] Meeker, Hop Culture [Puyallup, Washington Territory, 1883], opposite p. 27.)

fan-blast kilns. A large fan at the base of the new kilns was driven by a twelve- or fifteen-horsepower steam engine. This fan, the buzzing of which was heard in the neighborhood "at all hours," forced a volume of air around the heated furnace and the pipes and then upward, through the drying floor.[81] The idea was not Meeker's alone.[82] Likewise the separation of the kilns from the cooling and baling facilities was not solely his idea. In both cases, however, Meeker's book *Hop Culture*, published in 1883, undoubtedly set the standards that were imitated elsewhere.[83]

In the Pacific Northwest as in California, a boom in the construction of hop driers occurred in the 1890s. For example, the Snoqualmie Company in Washington, the "largest hop ranch in the world," expanded to include seventeen hop-cooling barns, and twenty-one driers, each twenty-four feet square. Growers in Polk County, Oregon, erected about twenty-five hop driers during the summer of 1894 alone.[84]

The idea that hops should be dried with drafts of forced hot air became increasingly common. One scheme called for the fireplace and stoves to be erected outside and adjoining the lower chamber of the kiln. A thick cast-iron pipe, about fourteen inches in diameter, proceeded through the fire two or three times before it passed, red hot, through the kiln wall. External air, sucked in by means of a fan and forced into the pipe to get heated, passed through the hops and left the kiln via the ventilator at the top. When curing was complete, the operator substituted cool air to accelerate the removal of the hops.[85]

The improvement of the heating devices did not radically alter the form of the buildings. The kilns in the hop yard of James Seavey, north of Springfield, Oregon, demonstrate three building forms used in the double kilns of the Willamette Valley. Apparently the oldest is the gable-roofed structure, second from the right. It may have had a cooling barn attached, in the manner of many large late-nineteenth-century hop houses. The two pairs of kilns on the left, each having a high hipped roof, are characterized by the bridge connecting their ventilators. By the time these three double kilns stood next to one another, a track system (at the far right) connected them to the cooling barn beyond. The new pair were somewhat imperfectly conceived, however, because the location of the stove chimneys—between the units—denies them the maximum draft possible. The last kiln solves this problem by moving the chim-

TINGED WITH GOLD

ney stacks to the outside wall. Further, the length of the bridge between the kilns is reduced to the minimum necessary.[86] The separation of the kilns from the cooling barn allowed either to be enlarged without affecting the other. The hipped double kiln forms, which became standardized to some degree, were characteristic of James Seavey's buildings. Seavey owned five hop fields in Oregon; the four pairs of kilns that he built on his Corvallis farm in 1911 had nearly the same form and arrangement.

While a certain degree of standardization occurred in the kilns of large growers, a considerable amount of variety was evident in the kilns used by smaller farmers. The financial resources of the grower, his ethnic background, and the materials at hand could all play a part in the choice of construction. This can be seen in one half of a double kiln built by Herman Bjerke, an emigrant from Norway, probably about 1910. In this instance, the structure is composed of a frame of peeled-log corner posts and sills, resting on fieldstone and oak-drum footings. The drier alone survives, because of its relatively isolated site, on the edge of a valley emptying into Coyote Creek; it is an important example of a hop structure constructed in a distinct ethnic fashion.

The hiatus in hop production during Prohibition did not completely stop the construction of driers and cooling barns, but it does seem to have slowed invention and improvement. Not until the early 1930s, particularly in Oregon, did growers revise and improve their methods of drying. Because any change in equipment was expensive, at first only the larger growers were responsible for experimentation. Hop buyers generally felt that the very large crop harvested in 1933 was of poor quality because of inadequate facilities. To provide a substitute for the old "air-dried" system, several growers hired engineers to assist them in designing fans in the ventilators of the kilns. As a result, while some of the new kilns pushed hot air through the hops from below, others both pushed air from below for a certain amount of time and then pulled the hot air through from fans above, to dry the top half of the kiln.[87] Such "vacuum dried" methods eliminated moisture, while presumably leaving behind more hop oils and resins. The engineers also claimed that this new type consumed less wood or oil per pound of hops.[88]

The Seavey Yards at Oregon City were one of the first to

5.38 James Seavey's kilns, north of Springfield, Oregon, c. 1885 to c. 1911. Three building campaigns are evident in this collection of kilns. (Lane County Historical Museum, Eugene, Oreg.)

5.39 Herman Bjerke's
kiln, Doane Road, Crow
vicinity, Lane County,
Oregon, c. 1910. The
remaining structure is
one of two kilns built
on the site. (Courtesy of
Maura Johnson.)

try this method. There, in 1933, the management installed a seven-foot airplane propeller, driven by a horizontally mounted seven-and-a-half-horsepower electric motor.[89] In 1935 the E. Clemens Horst Company installed similar vacuum fans in a new six-drier kiln on its ranch near Independence, Oregon. This was one of nine new kilns equipped by the Moore Dry Kiln Company, which also introduced new fan equipment in three of the old driers on the property.[90]

By the mid-1930s the increasing interest in forced-draft drying led many smaller growers to install electric fans in order to shorten drying time. Often existing kilns would be improved or new ones constructed as the power company extended its lines into a rural area. It became common to see motors mounted on the roof, either beside or above the ventilator, with a protective cover or roof over them. Manufacturers such as Sloper, Sturdevant, Moore, and Eastman became well known for their fans.

Electricity was also used for other purposes. As early as the mid-1920s, a few hop growers in the Willamette Valley began to

5.40 *Herman Bjerke's kiln, 1910, interior. The structure is composed of a frame of peeled-log corner posts and sills on a field-stone footing. (Courtesy of Maura Johnson.)*

experiment with electric pumps for irrigation of their fields, a technique that became more generally accepted in the next decade.[91] In the early 1930s Arch Sloper developed an electric hop baler with controls reminiscent of an automatic elevator, where operators needed only to press buttons to operate the machine.[92] In addition to these devices, growers made more use of automatic controlling and recording instruments, especially thermostats and humidistats.

The development of these increasingly sophisticated aids meant that specialists became more involved with design and construction on the hop farm. By 1933 the carpenter Homer Wood, who had erected at least sixty hop kilns in and around

5.41 *Twin fan installa-*
*tion, Oregon, c. 1935, by*
*the Moore Dry Kiln Com-*
*pany, (From C. J. Hurd,*
*"Fans in Hop Dryers,"*
Pacific Hop Grower *2,*
*no. 9 [January 1934]; 6.)*

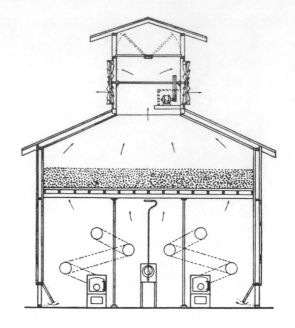

Independence, Oregon, was an unusual celebrity.[93] Wood's were frame kilns, although they were by no means meager undertakings. The two thirty-four-foot frame kilns that he built for the W. H. Walker ranch northeast of the city were equipped with electric service fans and had large storerooms connected above ground.[94] After Wood's time, however, few men earned a living by constructing such buildings in the traditional manner.

Architects and engineers became involved when commercial operations required a more sophisticated understanding, or when a knowledge of materials other than wood was necessary. For example, in 1933 Northwest Brogdex Company, a firm of mechanical engineers, and Francis H. Fassett, an architect, drew up the scheme for the hop kiln and warehouse of Lloyd L. Hughes, Incorporated, at Moxee City, Washington. The plans called for a boiler with a large tank of steam coils and a large blower fan, thermostatically controlled, which forced air through the hops at a low temperature. The concrete building, which was used for curing, baling, and storing hops, remains a prominent local monument, standing nearly empty. In 1935 the Portland architecture firm of Knighton and Howell provided plans for the five-story, 120-by-80-foot concrete hop warehouse erected by the Southern Pacific Railroad in Salem, Oregon, at a cost of one hundred thousand dollars.[95] In both

these cases, insurance costs were reportedly lower because of
the use of an incombustible material and the installation of
sprinkler systems.

Western technology also had a slight effect on the form of
new driers in New York State. A consortium of Washington
State and New York growers attempted to introduce modern,
large-scale production methods to the East by constructing a
farm complex about two miles north of Bridgewater on the east
side of State Route 8. The Oneida Chief Farms grew approximately 250 acres of hops for a few years, but the results of the
experiment simply were not worth the effort.[96] There would be
no renaissance of hop culture in New York in the twentieth
century.

After World War II some growers in the Pacific Northwest
adopted a completely new form of hop drier, designed for
modest-sized farms. This can be seen in a bulletin published by
the Agricultural Experiment Station at Oregon State College
in Corvallis in 1950. The drawings provided for a single building, twenty-eight by seventy-five feet long, including a furnace
room, a kiln, a cooler, a storage room, and a baling room. The
bulletin recommended a poured concrete foundation, concrete-
block walls and a metal monitor roof.[97] Corrugated steel also

*5.42   Lloyd L. Hughes,
Inc., hop kiln and ware-
house, Moxee City, Yak-
ima County, Washington,
1933. Northwest Brogdex
Company, mechanical
engineers, and Francis H.
Fassett, architect, were
responsible for this reinforced concrete structure.
(From "New Airblast
Kilns Erected at Moxee
City," Oregon Hop
Grower 1, no. 4 [July
1933]: 1.)*

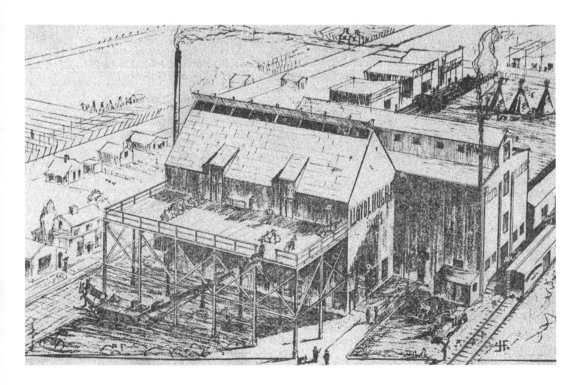

5.43 Oneida Chief Farms, near Bridgewater, Oneida County, New York, constructed c. 1940. The four outlying kilns connected to the cooling and baling barn in the center [originally constructed as a dairy barn] is a distinct arrangement transplanted from the Pacific Coast.

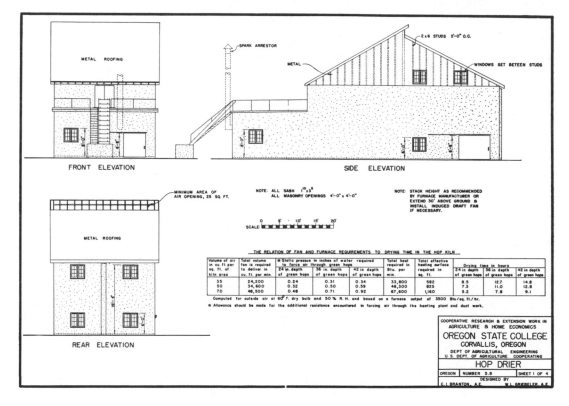

THE RELATION OF FAN AND FURNACE REQUIREMENTS TO DRYING TIME IN THE HOP KILN

| Volume of air in cu. ft. per sq. ft. of kiln area | Total volume fan is required to deliver in cu. ft. per min. | ⋈ Static pressure in inches of water required to force air through green hops | | | Total heat required in Btu. per min. | Total effective heating surface required in sq. ft. | Drying time in hours | | |
|---|---|---|---|---|---|---|---|---|---|
| | | 24 in. depth of green hops | 36 in. depth of green hops | 42 in. depth of green hops | | | 24 in depth of green hops | 36 in depth of green hops | 42 in depth of green hops |
| 35 | 24,200 | 0.24 | 0.31 | 0.34 | 33,800 | 582 | 8.5 | 12.7 | 14.8 |
| 50 | 34,600 | 0.32 | 0.50 | 0.59 | 48,300 | 825 | 7.3 | 11.0 | 12.8 |
| 70 | 48,500 | 0.48 | 0.71 | 0.92 | 67,600 | 1,160 | 5.2 | 7.8 | 9.1 |

Computed for outside air at 60° F. dry bulb and 50 % R. H. and based on a furnace output of 3500 Btu/sq. ft./hr.

⋈ Allowance should be made for the additional resistance encountered in forcing air through the heating plant and duct work.

COOPERATIVE RESEARCH & EXTENSION WORK IN AGRICULTURE & HOME ECONOMICS

OREGON STATE COLLEGE
CORVALLIS, OREGON

DEPT OF AGRICULTURAL ENGINEERING
U. S. DEPT. OF AGRICULTURE COOPERATING

HOP DRIER

| OREGON | NUMBER 5.8 | SHEET 1 OF 4 |

DESIGNED BY
C. I. BRANTON, A.E.     W.I. GRIEBELER, A.E.

5.44 Plans for hop drier, Oregon Farms, Agricultural Experiment Station at Oregon State College in Corvallis, 1950. The drawings provided for a single building, 28 by 75 feet long. Ventilation is through a curtain at the peak of the shed roofs in the manner of the Hughes kiln (figure 5.42), giving the drying building a modern profile. (From C. Ivan Branton, A Hop Drier for Oregon Farms [Corvallis, Oreg.: Agricultural Experiment Station, Oregon State College, 1950], sheets 1–4.)

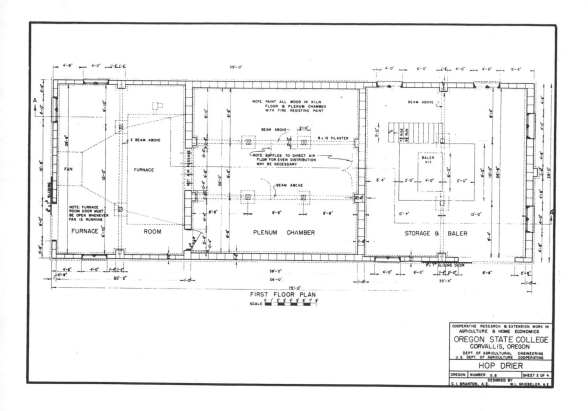

NOTE PAINT ALL WOOD IN KILN
FLOOR & PLENUM CHAMBER
WITH FIRE RESISTING PAINT

BEAM ABOVE

8 x 16 PILASTER

NOTE: BAFFLES TO DIRECT AIR
FLOW FOR EVEN DISTRIBUTION
MAY BE NECESSARY

BEAM ABOVE

BEAM ABOVE

7 RISE
30 RUN

BALER
PIT

FAN

FURNACE

NOTE: FURNACE
ROOM DOOR MUST
BE OPEN WHENEVER
FAN IS RUNNING

FURNACE          ROOM          PLENUM CHAMBER          STORAGE & BALER

SLIDING DOOR

I BEAM ABOVE

## FIRST FLOOR PLAN
SCALE

COOPERATIVE RESEARCH & EXTENSION WORK IN
AGRICULTURE & HOME ECONOMICS

### OREGON STATE COLLEGE
CORVALLIS, OREGON
DEPT. OF AGRICULTURAL ENGINEERING
U.S. DEPT. OF AGRICULTURE COOPERATING

#### HOP DRIER

| OREGON | NUMBER 5.8 | SHEET 2 OF 4 |
| --- | --- | --- |

DESIGNED BY
C. I. BRANTON, A.E.    W. L. GRIEBELER, A.E.

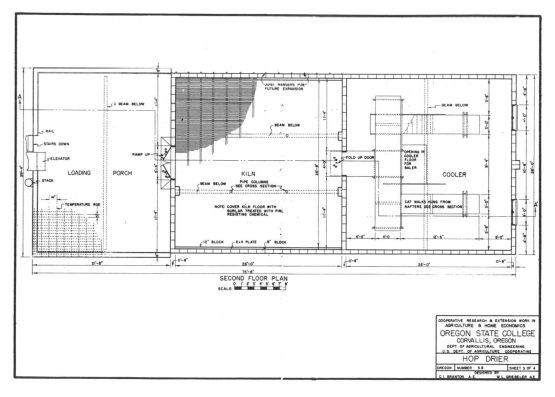

JOIST HANGERS FOR
FUTURE EXPANSION

I BEAM BELOW

BEAM BELOW

BEAM BELOW

RAIL

STAIRS DOWN

ELEVATOR

RAMP UP

FOLD UP DOOR

OPENING IN
COOLER
FLOOR
FOR
BALER

STACK

LOADING    PORCH          KILN          COOLER

TEMPERATURE ROD

BEAM BELOW

PIPE COLUMNS
SEE CROSS SECTION

NOTE: COVER KILN FLOOR WITH
BURLAP TREATED WITH FIRE
RESISTING CHEMICAL

CAT WALKS HUNG FROM
RAFTERS SEE CROSS SECTION

12" BLOCK    2 x 4 PLATE    8" BLOCK

## SECOND FLOOR PLAN
SCALE:

COOPERATIVE RESEARCH & EXTENSION WORK IN
AGRICULTURE & HOME ECONOMICS

### OREGON STATE COLLEGE
CORVALLIS, OREGON
DEPT. OF AGRICULTURAL ENGINEERING
U.S. DEPT. OF AGRICULTURE COOPERATING

#### HOP DRIER

| OREGON | NUMBER 5.8 | SHEET 3 OF 4 |
| --- | --- | --- |

DESIGNED BY
C. I. BRANTON A.E.    W. L. GRIEBELER A.E.

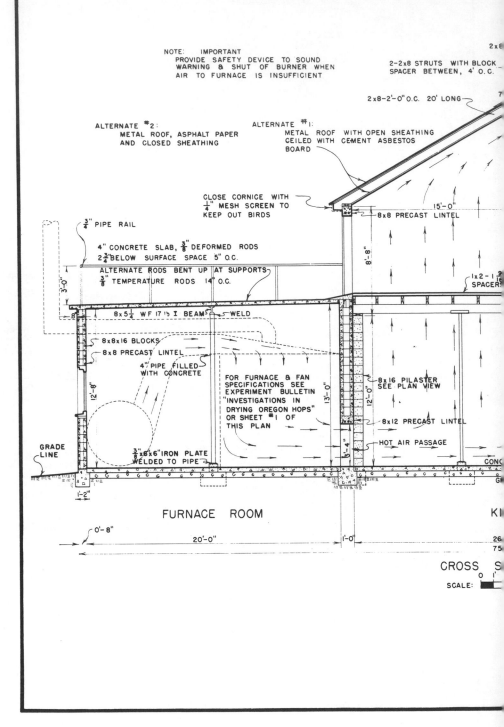

NOTE:   IMPORTANT
PROVIDE SAFETY DEVICE TO SOUND
WARNING & SHUT OF BURNER WHEN
AIR TO FURNACE IS INSUFFICIENT

2x8

2-2x8 STRUTS WITH BLOCK
SPACER BETWEEN, 4' O.C.

2x8-2'-0" O.C.  20' LONG

ALTERNATE #2:
METAL ROOF, ASPHALT PAPER
AND CLOSED SHEATHING

ALTERNATE #1:
METAL ROOF WITH OPEN SHEATHING
CEILED WITH CEMENT ASBESTOS
BOARD

CLOSE CORNICE WITH
¼" MESH SCREEN TO
KEEP OUT BIRDS

15'-0"

8x8 PRECAST LINTEL

¾" PIPE RAIL

4" CONCRETE SLAB, ⅜" DEFORMED RODS
2 ¾" BELOW SURFACE SPACE 5" O.C.
ALTERNATE RODS BENT UP AT SUPPORTS
⅜" TEMPERATURE RODS 14" O.C.

8'-8"

1x2-1
SPACER

3'-0"

8x5¼ WF 17 ᴵᵇ I BEAMS      WELD

8x8x16 BLOCKS

8x8 PRECAST LINTEL

4" PIPE FILLED
WITH CONCRETE

FOR FURNACE & FAN
SPECIFICATIONS SEE
EXPERIMENT BULLETIN
"INVESTIGATIONS IN
DRYING OREGON HOPS"
OR SHEET #1 OF
THIS PLAN

8x16 PILASTER
SEE PLAN VIEW

8x12 PRECAST LINTEL

HOT AIR PASSAGE

12'-8"

13'-0"

12'-0"

GRADE
LINE

¾"x6x6" IRON PLATE
WELDED TO PIPE

4"

CONC

1'-2"

FURNACE ROOM

K

0'-8"

20'-0"

1'-0'

26
75

CROSS S

SCALE:

0   1

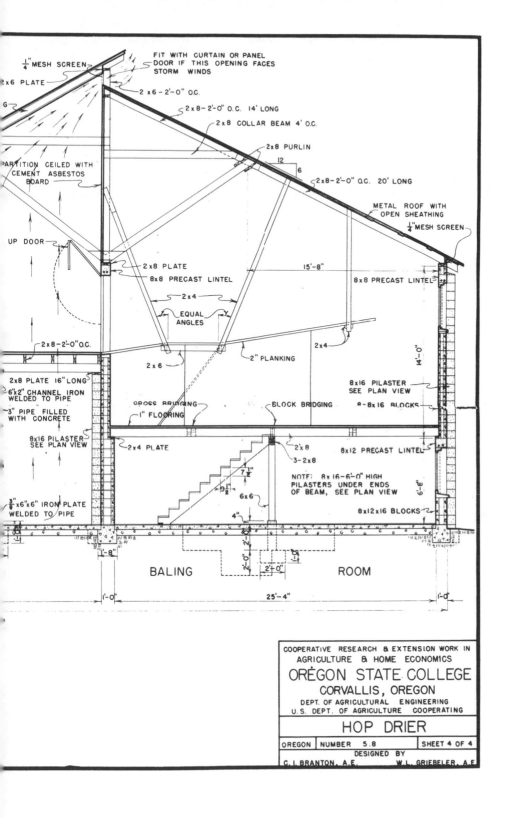

FIT WITH CURTAIN OR PANEL DOOR IF THIS OPENING FACES STORM WINDS

¼" MESH SCREEN

2 x 6 PLATE

2 x 6 - 2'-0" O.C.

2 x 8 - 2'-0" O.C. 14' LONG

2 x 8 COLLAR BEAM 4' O.C.

2 x 8 PURLIN

12

6

2 x 8 - 2'-0" O.C. 20' LONG

PARTITION CEILED WITH CEMENT ASBESTOS BOARD

METAL ROOF WITH OPEN SHEATHING

¼" MESH SCREEN

UP DOOR

2 x 8 PLATE

8 x 8 PRECAST LINTEL

15'-8"

8 x 8 PRECAST LINTEL

2 x 4

EQUAL ANGLES

2 x 4

14'-0"

2 x 8 - 2'-0" O.C.

2 x 6

2" PLANKING

2 x 8 PLATE 16" LONG

6"x 2" CHANNEL IRON WELDED TO PIPE

3" PIPE FILLED WITH CONCRETE

8 x 16 PILASTER SEE PLAN VIEW

8 x 16 PILASTER SEE PLAN VIEW

8 - 8 x 16 BLOCKS

CROSS BRIDGING

BLOCK BRIDGING

1" FLOORING

2 x 4 PLATE

2 x 8

3 - 2 x 8

8 x 12 PRECAST LINTEL

⅜"x 6"x 6" IRON PLATE WELDED TO PIPE

7½"

9½"

6 x 6

4"

NOTE: 8 x 16 - 6'-0" HIGH PILASTERS UNDER ENDS OF BEAM, SEE PLAN VIEW

6'-8"

8 x 12 x 16 BLOCKS

1'-8"

2'-0"

2'-0"

BALING

ROOM

1'-0"

25'-4"

1'-0"

COOPERATIVE RESEARCH & EXTENSION WORK IN
AGRICULTURE & HOME ECONOMICS
OREGON STATE COLLEGE
CORVALLIS, OREGON
DEPT. OF AGRICULTURAL ENGINEERING
U.S. DEPT. OF AGRICULTURE COOPERATING

HOP DRIER

| OREGON | NUMBER 5.8 | SHEET 4 OF 4 |

DESIGNED BY
C. I. BRANTON, A.E.        W. L. GRIEBELER, A.E.

began to be more frequently used for the exterior wall sheathing after World War II. These materials are no more maintenance free than any others, as can be seen in the Yakima Valley today.

The evolution from the banked hop kiln to the four-room hop house was a result of several improvements, primarily the introduction of the stove and thin-wall frame construction. This synthesis occurred in New York State, at the start of its rise to prominence as a major hop-producing region. The evolution continued in Wisconsin, where growers first used the balloon frame, and developed a model in which the drying function was separated from the cooling barn. This may be seen as a pivot point, for the Pacific Coast growers, having watched their eastern brethren, made use of this idea as they attempted to mechanize the drying process. West Coast growers embraced not only the concept of using carts and conveyors to move the hops, but also ideas that involved multiple kilns, fan blasting, and vacuum drying to expedite curing. Of course, experimentation goes on. Recent studies in the use of solar collectors, sponsored by the United States Departments of Agriculture and Energy, may point the way into the future.[98]

TINGED WITH GOLD

# ❧ Conclusion ❧

T his work sought to explain one aspect of the agrarian landscape: hop culture. As the preceding chapters indicate, growers raised hops in specific areas of the country and in each, during distinct time periods.

The historical development can be divided into three periods. The first begins at the end of the eighteenth century, with the introduction of commercial hop growing in Massachusetts and continued until the 1840s, when New York began to dominate the domestic market. Growers in the United States were aware of English precedents and practice, but few parallels can be drawn. Whereas farmers in southeastern England turned to London for additional labor, in the fields of New England and New York hop growers hired home pickers from the immediate region. Young farm women found hop picking a means of earning money and enjoying companionship in the fields, an agreeable alternative to routine chores in the days before mill towns offered another way of life. Likewise, the ideas that lay behind the technologically more advanced hop kilns of Great Britain were only dimly understood. Early growers in the United States did not readily adopt the latest English methods or inventions. They burned wood instead of charcoal because it was inexpensive and plentiful. The banked kiln, rather than the more efficient oast form, continued in use in the early-nineteenth-century hop-growing regions of New England and New York. The invention of the stove was adopted only gradually, and the conically shaped draft roof came to be an image associated with the poled fields at least three or four decades after it had first been employed in Kent.

The second period, from the mid-nineteenth century to the advent of Prohibition, was largely a time of robust expansion and growth, when the United States became recognized worldwide as a major producer. The booming culture in New York State was complemented by the meteoric rise in Wisconsin, while the planting of the first roots in California, Oregon, and Washington made hop culture a nationwide concern. City pickers entered the fields for the first time, and ethnic groups became identifiable cohorts, although again women outnumbered men. As the scale of operations increased, financial concerns become paramount, drawing some farmers into becoming dealers. Dealers in turn become brokers, bankers, and financiers. Growers who felt that their interests and investments were threatened organized in an attempt to solve their common problems. In the fields technological inventions were adopted. The trellis superseded the old-fashioned pole system in most areas by the end of the century. The masonry kiln, banked into the side of the hill, developed into the freestanding, braced-frame, four-room hop house, which was built with a balloon frame in the Midwest. It evolved, in turn, from a single building into two, the kiln and the cooling barn, an arrangement introduced on the West Coast almost at the same time as hop growing was recognized there as a major agricultural industry.

The third phase moved the industry through Prohibition and continues to the present, primarily in the Pacific Northwest. This period has seen the demise of the small growers as the problems associated with a fluctuating demand for hops in the United States and abroad meant that the return on one's investment was more than ever a matter determined by the international market. The most obvious change in the fields was the reduction of numbers of workers, who have been largely replaced by the mechanical processor. Modern-day hop driers have become mechanical tools; they symbolize a different period of hop culture and have completely lost their role in the community as entertainment centers. In large fields the kiln and the cooling barn were separated into four or even five buildings, each with its own function.

A number of events in this history should be underscored, and a number of important people should not be forgotten. The Tacoma anti-Chinese riots and the conflict at Wheatland, California, are but two outstanding examples. To social historians interested in ethnicity and the plight of field workers, these controversies should have special meaning. Growers like Cooledge, Blanchard, Clarke, Pilot, Durst, Meeker, and the Seavey brothers became legendary in their own time. Through their promotion and enterprise they provided a significant amount of the impetus for others to become involved. Of this list, only the pioneer Meeker has received any attention, not for his role as a hop grower but rather for his association with the Oregon Trail. The dealers and brewers also deserve attention. Matthew Vassar's role is paramount in New York State; others remain to be discovered. The story of hop growing in the United States will not be complete until their careers are carefully investigated.

Biographical studies of carpenter-builders, such as Homer Wood, are important for understanding the character of the hop buildings in a particular region. On the other hand, the role of the New York carpenter-builder W. A. Brooks in providing Wisconsin growers with the latest hop-house building ideas in 1868 lends some insight about the ways in which inventions from one area of the country were transplanted to another. How tantalizing would be the prospect of more information on the Asian or Scandinavian builders of the Pacific Coast!

Related to this is the necessity for more surveys to identify specific buildings or groups of structures, particularly those near major market centers. In New York State, for example, the settlement of Phoenix, all but completely gone, should not be allowed to disappear without documentation. More important, the existing hop buildings in the Cooperstown area should be conserved and their future safeguarded. The Historic American Engineering Record recorded one of the surviving Seavey complexes in Oregon in anticipation of its demise, but the conservation of the complex has never been publicly discussed. The problem is all the more complicated because, as with many rural structures, the site must be conserved if one is to understand how harvesting activities took place. The orientation is especially important. Only two hop houses have been "saved." Americana Village, south of Madison, New York, and the Genesee Country Village, near Mumford, New York, each have taken the step of conserving a hop house by relocating it.

Having told the story of one class of "buildings behind the farmhouse," it should be apparent that agricultural structures are often much more sophisticated than is generally acknowledged. Far from being manifestations of some folk tradition, these buildings are often the result of carefully calculated commercial motives. The activities that went on in the field and those that were performed in the structure are so closely related that neither can be ignored without hampering the understanding of the other. Further, if hop buildings are any indication, the processes are more important than the product in determining the form and arrangement of an agricultural structure. From one side of the country to the other, these principles should be applied in a reevaluation of rural vernacular buildings. Let every structure present its story. This broadening awareness of design and construction can only provide for a better understanding and appreciation of the built environment and lend strength to our efforts to conserve it.

# APPENDIX

One of the most complete descriptions ever to be published of post–Civil War hop house construction was that written by Sereno Edwards Todd.[1] Born and raised in Lake Ridge, Tompkins County, New York, Todd received a sporadic education at a number of schools and academies in the area before trying his hand at farming and domestic construction. However, by the late 1850s writing had become his chief occupation, and he contributed to various agricultural journals many articles on both agriculture and farm construction. In 1865 Todd moved to New York City, where he was an associate editor of the *American Agriculturalist* from 1865 to 1866, the editor of the "Agricultural and Livestock Department" of the *New York Observer* in the late 1860s, an editor for *Hearth and Home*, and an agricultural editor of the *New York Tribune* by the mid 1870s. He also held a position on the *New York Herald* and edited the *Practical Farmer* until he retired in 1881.

## Hints about Hop Houses

A hop-kiln twenty feet square answers extremely well for the purposes of a yard of from four to ten acres, although if used for the latter number, with a full crop, the picking would have to be regulated according to the capacity of the kiln, especially in the beginning, when, owing to the quantity of moisture in the green hops, the process of drying is much less rapid than in later stages of picking. A kiln of the size mentioned may have the rafters twenty feet long, starting from each of the four sides, and approaching within three feet of each other, in the center, thus leaving an opening three feet square for a ventilator. This opening may be covered by a revolving cowl, or by a fixed slatted ventilator. Upon the ground level there should be numerous openings, say six or eight in all, upon different sides of the kiln, and these should be fitted with slide, so as to be closed at pleasure. It is important that the area of these openings should, in the aggregate, at least equal—they had better exceed—the area of the opening left in the roof for ventilation. Should a violent wind prevail at any time during drying, so as to drive the fire within the kiln, the slides upon the windward side may be closed. As soon as the hops upon the kiln are dried, all the slides should be closed, as so to stop the draft of air through them, for at that stage it can do no good.

A chimney may be constructed inside the kiln, or a stovepipe may be carried up outside. A chimney economizes the heat, and, in the long run, if properly constructed, is the safer of the two. Up to a point at least a foot above the hops it should be eight inches thick, beyond that four inches will answer. Where the hops would live against the chimney, it should be enclosed by a wooden partition, four inches from the chimney, so as to keep the hops from contact with the bricks.

It is now considered that a space of fourteen feet between the cloth and the ground is desirable. In the old kilns it was very common to have the space only eight or nine feet, and these kilns worked well, but required much care. The advantages of a high kiln are that there is no danger of scorching the hops, that the heat is more equally diffused over the entire surface, and the fumes of the sulphur come more readily and certainly in contact with every portion of the hops upon the cloth. Some have raised the floor to sixteen and even eighteen feet, but experience does not commend that practice.

The ground floor may consist of well-rammed earth. The kiln floor should consist of slats. These are generally one and a half inches square, and placed one and a half inches apart. Slats, one and a half by two inches, and two inches apart, laid with the one and a half inch face up, are preferred by many. The joint should be very smooth, and nailed so as to resist warping, or the cloth will suffer damage. These slats should run in the same direction as you wish to remove the hops after drying. This floor is covered by a cloth or carpet, manufactured for the purpose. The sides, as high as the plates, should be thoroughly plastered to retain the heat, but the ceiling above the plates may be left unfinished; it is not desirable to have plaster *over* the hops. Both the fire room and the drying room should be well-lighted by moderate-sized windows, so that the entire process may be carefully observed. Too much window, however, would cause a waste of heat. The door of the fire room, and the doors of the drying room, should each be finished with a single pane of glass, so that the condition of the interior may be noted without opening the door. Attached to one side of the building there should be a large and strong platform, to receive the hops in sacks, and a door should open into the drying room.

A kiln, such as is described, would constitute only a portion of a hop-house, for it has no room for storage. The floor of the store room may be on a level with, or several feet lower than the floor of the drying room, and should be large enough to contain the entire crop, for there is always some loss in weight and risk of heating in the bale, when the hops are pressed in less than three or four weeks after drying. It is in the store room that the process of curing is completed. If the hops *must* be packed directly from the kiln, they must also be over-dried to fit them for such packing. This floor should be jointed, or seasoned stuff, and as smooth as possible. At a convenient place, with reference to the press on the lower floor, an opening, twelve or more inches square, should be left, in order to feed the press as the time of packing. The windows of the store room should be darkened while the hops are stored, as exposure to light is calculated to impair their brightness. The ground floor of the store room may be fitted and used for any purpose of storage, and will always be found useful on a farm. Double kilns are frequently used for large yards, but very often are less manageable than those constructed singly, as they are less controllable in the matter of draft. A double kiln is made by running a partition so as to make two fire rooms and two kilns under one roof, so that fresh hops may be laid upon one before those partially dried are removed from the other.

# NOTES

## Introduction

1. Eric Arthur and Dudley Whitney, *The Barn: A Vanishing Landmark in North America* (Toronto: McClelland and Stewart, 1972).

2. Carl Sauer, "The Morphology of the Landscape," *University of California Publications in Geography* 2, no. 2 (1925): 19–54.

3. Fred Kniffen, "Louisiana House Types," *Annals of the Association of American Geographers* 26 (1936): 173–93; and "Folk Housing: Key to Diffusion," *Annals of the Association of American Geographers* 55 (1965): 549–77.

4. Henry Glassie, *Pattern in the Material Folk Culture of the Eastern United States* (Philadelphia: University of Pennsylvania Press, 1968).

5. Ibid., 160. The large barns in this area were presumably related to English oast houses.

6. See, for example, Henry Glassie, *Folk Housing in Middle Virginia* (Knoxville: University of Tennessee Press, 1975).

7. Otis W. Freeman, "Hop Industry of the Pacific Coast States," *Economic Geography* 19, no. 3 (April 1936): 155–63; Edgar M. Hadley, "Hops in the United States," *Geographical Bulletin* 2, no. 1 (April 1971): 37–51; Paul H. Landis, "The Hop Industry: A Social and Eco-

nomic Problem," *Economic Geography* 22, no. 2 (January 1939): 85–94; Elbert Miller and Richard M. Highsmith, "The Hop Industry of the Pacific Coast," *Journal of Geography* 49, no. 2 (1950): 63–77; James Jerome Parsons, Jr., "The California Hop Industry: Its Eighty Years of Development and Expansion" (Master's thesis, University of California at Berkeley, 1939); Thomas Rumney, "Agricultural Production Locational Stability: Hops in New York State During the Nineteenth Century," *Kansas Geographer* 18 (1983): 5–16; Thomas Rumney, "The Hops Boom in Nineteenth Century Vermont," *Vermont History* 56, no. 1 (Winter 1988): 36–41.

8. Charles F. Calkins and William G. Laatsch, "The Hop Houses of Waukesha County, Wisconsin," *Pioneer America* 9, no. 2 (December 1977): 180–207.

9. Allen G. Noble, *Wood, Brick, and Stone: The North American Settlement Landscape*, vol. 2, *Barns and Farm Structures* (Amherst: University of Massachusetts Press, 1984), 100–102.

10. John Fitchen, *The New World Dutch Barn* (Syracuse, N.Y.: Syracuse University Press, 1968); Theodore H. M. Prudon, "The Dutch Barn in America: Survival of a Medieval Struc-

tural Frame," *New York Folklore* 2, nos. 3–4 (1976): 122–42, reprinted in *Common Places: Readings in American Vernacular Architecture*, edited by Dell Upton and John Michael Vlach (Athens: University of Georgia Press, 1986), 204–14.

## 1. The History of Hop Growing in the United States

1. Edward Lance, *The Hop Farmer; or, A Complete Account of Hop Culture, Embracing Its History, Laws, and Uses: A Theoretical and Practical Inquiry into an Improved Method of Culture, Founded on Scientific Principles* (London: Joseph Rogerson, 1838), 1–2. Hops were grown throughout most of Europe by the early sixteenth century. The culture was introduced from the Low Countries to England about 1525, and by 1603 several statutes had been passed regulating the inspection and curing of hops.

2. John P. Arnold and Frank Penman, *History of the Brewing Industry and Brewing Science in America, Prepared as Part of a Memorial to the Pioneers of American Brewing Science, Dr. John E. Siebel and Anton Schwarz* (Chicago: G. L. Peterson, 1933), 33.

3. The Spanish settled in the New World earlier, but there is no indication that they preferred beer.

4. Edmund Bailey O'Callaghan, *History of New Netherland; or, New York Under the Dutch* (New York: D. Appleton, 1846), 156; Edmund Bailey O'Callaghan, *Laws and Ordinances of New Netherland, 1638–1674* (Albany, N.Y.: Weed, Parsons, 1868), 39, 71, 80.

5. David Pietersz De Vries, *Voyages from Holland to America, A.D. 1632 to 1644*, trans. Henry C. Murphy (New York: James Lenox, 1853), 157.

6. Edmund Bailey O'Callaghan, ed., *Documents Relating to the Colonial History of the State of New York* (Albany, N.Y.: Weed, Par-

sons, 1856), 1: 634–35. The same general pattern of agrarian development was evident in the Dutch settlements on the upper Hudson. O'Callaghan, *History of New Netherland*, 388–89; Charlotte Wilcoxen, *Seventeenth Century Albany: A Dutch Profile* (Albany, N.Y.: Albany Institute of History and Art, 1981), 51.

7. Lyman Carrier, *The Beginnings of Agriculture in America* (New York: McGraw-Hill, 1923), 252. Wild hops were noted as "fair and large" in a pamphlet describing Virginia, published in 1649. See Stanley Brown, *Brewed in America: A History of Beer and Ale in the United States* (Boston: Little, Brown, 1962), 10–12; "Hops: Otsego County, N.Y.," *Working Farmer* 7, no. 3 (April 1855): 45; James F. W. Johnston, "The Hop and Its Substitutes, from 'The Chemistry of Common Life,'" *Working Farmer* 7, no. 4 (June 1855): 91.

8. Karen Ordahl Kupperman, "Climate and Mastery of the Wilderness in Seventeenth-Century New England," in *Seventeenth Century New England*, ed. David D. Hall and David Greyson Allen (Boston: Colonial Society of Massachusetts, 1984), 3–37; Howard S. Russell, *A Long, Deep Furrow: Three Centuries of Farming in New England* (Hanover, N.H.: University Press of New England, 1976), 134, 298; Brown, *Brewed in America*, 68.

9. William Blanchard, "Hop Culture," *Prairie Farmer*, n.s., 19, no. 19 (May 1867): 315; Samuel Adams Drake, *History of Middlesex County, Massachusetts, Containing Carefully Prepared Histories of Every City and Town in the County* (Boston: Estes and Lauriat, 1880), 2:509.

10. "Culture of Hops in Massachusetts," *Yankee Farmer and Newsletter*, July 28, 1838.

11. Rudolphus Dickinson, *A Geographical and Statistical View of Massachusetts Proper* (Greenfield, Mass.: Denio and Phelps, 1813), 52–53.

12. *Eighty Years' Progress of the United States: A Family Record of American Industry, Energy, and Enterprise* (Hartford, Conn.: L. Stebbins, 1868), 1:88. Similar standards were

regularly established to regulate the quality of hides, lumber, and beef in many Eastern states.

13. Massachusetts Valuation Returns, 1767, 1801, 1811 (in the Archives of the Massachusetts State House, Boston, Massachusetts), as noted in Percy Wells Bidwell and John I. Falconer, *History of Agriculture in the Northern United States* (Washington, D.C.: Carnegie Institution, 1925), 243.

14. Harold Fisher Wilson, *The Hill Country of Northern New England: Its Social and Economic History, 1790–1930* (New York: Columbia University Press, 1936), 23–26.

15. *History of Bedford, New-Hampshire, Being Statistics, Compiled on the Occasion of the One Hundredth Anniversary of the Incorporation of the Town* (Boston: Alfred Mudge, 1851), 203–4. The 1840 census shows Hillsborough County to have produced 174,196 pounds of the 243,425 pounds that the entire state harvested.

16. William P. Colburn, *The History of Milford* (Concord, N.H.: Rumford Press, 1901), 321, 390; "Culture of Hops," *Farmer's Monthly Visitor* 1, no. 3 (March 1839): 34; "Hop Cultivation," *Farmer's Monthly Visitor* 1, no. 6 (June 1839): 83. See also David F. Secomb, *History of the Town of Amherst, Hillsborough County, New Hampshire* (Concord, N.H.: Evans, Sleeper and Woodbury, 1883), 147; J. D. B. DeBow, supt., *The Seventh Census of the United States: 1850* (Washington: Robert Armstrong, 1853), 27; Alonzo J. Fogg, comp., *The Statistics and Gazetteer of New Hampshire* (Concord, N.H.: D. L. Guernsey, 1874), 437; James R. Jackson, *History of Littleton, New Hampshire, in Three Volumes* (Cambridge, Mass.: University Press, 1905), 334; Francis A. Walker and Charles W. Seaton, supts., *Report on the Productions of Agriculture as Returned at the Tenth Census (June 1, 1880)* (Washington, D.C.: U.S. Government Printing Office, 1883), 297.

17. Clarence Albert Day, "A History of Maine Agriculture, 1604–1860," *University of Maine Bulletin* 56, no. 11 (April 1954): 126, 130; Clar-

ence Albert Day, *Farming in Maine, 1860–1940* (Orono: University of Maine Press, 1963), 2, 19, 20.

18. Thoms Rumney, "The Hops Boom in Nineteenth Century Vermont," *Vermont History* 56, no. 1 (Winter 1988): 36–41.

19. J. D. B. DeBow, ed., *Statistical View of the United States . . . Being a Compendium of the Seventh Census* (Washington, D.C.: A. O. P. Nicholson, 1854), 172.

20. "Culture of Hops," *Working Farmer* 7, no. 3 (April 1855): 70–71. Vermont also had an inspector of hops appointed by the governor, although the post was later abolished. See Juar E. Jameson, "No. VII," in *Hop Culture: Practical Details from the Selection and Preparation of the Soil, and Setting and Cultivation of the Plants, to Picking, Drying, Pressing, and Marketing the Crop*, ed. Andrew S. Fuller (New York: Orange Judd, 1866), 32; DeBow, *Seventh Census*, 42; Walker and Seaton, *Report on Agriculture*, 319. It should be noted that the comment made by Ulysses Prentice Hedrick in *A History of Agriculture in the State of New York* (Albany: New York State Agricultural Society, 1933), 156, that "in the early half of the nineteenth century Vermont produced seven-eighths of the entire crop in the United States" is completely without foundation. Hedrick provides no supporting documentation, and there is ample evidence that Massachusetts was the acknowledged leader.

21. Ezra Meeker, "Old Time Prices," *Hop Culture in the United States, Being a Practical Treatise on Hop Growing in Washington Territory* (Puyallup, Washington Territory: E. Meeker, 1883), 121.

22. "To Hop Growers," *New England Farmer and Horticultural Register* 21, no. 12 (September 1842): 89.

23. As late as 1834, New York growers were comparing themselves to their eastern colleagues. "Gathering and Curing Hops," *Cultivator* 1, no. 1 (March 1834): 11.

24. Thomas Wilson, *Picture of Philadelphia,*

for 1824, Containing the "Picture of Phila-delphia, for 1811, by James Mease, M. D.," with All Its Improvements Since That Period (Phila-delphia: Thomas Town, 1823), 78; John Skinner, ed., "The Hop," *American Farmer: Rural Economy, Internal Improvements, Prices Cur-rent* 2, no. 39 (December 1820): 1; "On the Rais-ing of Hops," *Transactions*, 152.

25. The earliest historical review of the growth of the industry in upstate New York was written by W. A. Lawrence. "Hop Culture in the State of New York" appeared in Meeker's *Hop Culture in the United States*. Every author who has written on the topic subsequently has accepted this account. See also *Madison County Times*, September 21, 1878.

26. John E. Smith, ed., *A Descriptive and Biographical Record of Madison County, New York* (Boston: Boston History Company, 1899), 122–23; Luna M. Hammond Whitney, *History of Madison County, State of New York* (Syracuse, N.Y.: Truair, Smith, 1872), 600, 610, 284; Ezra Leland, "Hops," in *Transactions of the New-York State Agricultural Society, Together with an Abstract of the Proceedings of the County Agricultural Societies, and the American In-stitute (of the City of New York)*, vol. IV, 1844 (Albany, N.Y.: E. Mack, 1845), 447–48.

27. *Revised Statutes of New York* (1829), 1:565. Growers in Madison, Augusta, and Eaton townships, led by Amos Maynard, almost imme-diately petitioned the legislature for changes. *Journal of the Assembly of the State of New York at Their Forty-third Session* (Albany: J. Buel, 1820), 50, 406.

28. Amos O. Osborn, "Sangerfield's History," *Waterville Times and Hop Reporter*, October 1, 1886.

29. *New York Farmer and Horticultural Re-pository* 2, no. 2 (February 1829): 47.

30. "The Culture of Hops," *Genesee Farmer and Gardener's Journal* 4, no. 1 (January 1834): 3.

31. See, for example, "The Hop Culture," *Cultivator* 2, no. 12 (February 1836): 190; "Hop Culture," *Genesee Farmer* 7, no. 15 (April

1837): 120; and "Hops," *Genesee Farmer* 7, no. 42 (October 1837): 329–30.

32. DeBow, *Statistical View of the United States*, 172.

33. "Hops: Great Yield," *Cultivator* 10, no. 1 (January 1843): 11.

34. Francis A. Walker, supt., *Compendium of the Tenth Census (June 1, 1880), Compiled Pur-suant to an Act of Congress Approved August 7, 1882* (Washington, D.C.: U.S. Government Printing Office, 1885), part 1, 798–801.

35. Meeker, *Hop Culture in the United States*, 110. In 1879 William P. Locke was con-sidered "the largest grower in America," ac-cording to the *Waterville Times*, but apparently he slipped to second place in the region shortly thereafter. ("Hops! Who Handles Them? Our Dealers!" *Waterville Times*, April 24, 1879). The largest field in the early twentieth century was that of James F. Clark, in Otsego County, covering 150 acres. Herbert Myrick, *The Hop: Its Culture and Cure, Marketing and Manufac-ture* (New York: Orange Judd, 1899), 8; "Scenes in the Hop Yards," *Grip's Valley Gazette* 1, no. 2 (October 1893): 2.

36. Walker and Seaton, *Report on Agricul-ture*, 257, 299. The average yield per acre of the top six hop-producing counties in New York State was 596 pounds, whereas the comparable figure for California was 1,213 pounds.

37. E. W. Brewster, "Notice of the Cultiva-tor," *Prairie Farmer* 3, no. 5 (May 1843): 102.

38. A modest amount of hop raising was con-ducted in northern and southern Illinois and Iowa as settlers who came from New England and New York tried their hand. W.E.S., "Hops," *Prairie Farmer*, n.s., 17, no. 6 (Febru-ary 1866): 83; "Hops and Their Culture," *Prairie Farmer*, n.s., 17, no. 10 (March 1866): 145; "Joshua," "Iowa Legislation About Hop-Growing," *Prairie Farmer*, n.s. 21, no. 20 (May 1868): 315; "Hop Growing in Iowa," *Country Gentleman* 23, no. 11 (March 1864): 170.

39. "Hop Growing in the West," *Prairie Farmer*, n.s., 17, no. 2 (January 1866): 24; "Profit from Hops," *Prairie Farmer*, n.s., 19,

no. 2 (January 1867): 19; "Hops in the West," *Prairie Farmer*, n.s., 19, no. 5 (February 1867): 73; "The Wild Hop," *Country Gentleman* 20, no. 2 (November 1862): 321.

40. Benjamin Horace Hibberd, "The History of Agriculture in Dane County Wisconsin" (Ph.D. dissertation, University of Wisconsin, Madison, 1902), 149.

41. *Portrait and Biographical Record of Waukesha County, Wisconsin* (Chicago: Excelsior, 1894), 199–200.

42. DeBow, *Seventh Census*, 932.

43. Andrew Jackson Turner, "The History of Fort Winnebago," *Collections of the State Historical Society of Wisconsin* (Madison, Wis.: Democratic Printing Company, 1898), 14:79–80.

44. *Transactions of the Wisconsin State Agricultural Society* (Madison: Atwood and Rublee, 1868), 7:37; *The History of Sauk County, Wisconsin* (Chicago: Western Historical Company, 1880), 619. "Harvey Canfield, Benjamin Colton and Mr. Cottington are believed to have been the pioneers in the business in the county." "Hops and the Panic of 1868," *History of Sauk County*, 366.

45. Hibberd, "Agriculture in Dane County," 149.

46. "Culture of Hops," *Wisconsin Farmer, and Northwestern Cultivator* 15, no. 7 (July 1863): 248–49; "Method of Cultivating Hops," *Wisconsin Farmer, and Northwestern Cultivator* 17, no. 6 (June 1865): 163–64.

47. "The Hop Interest," *Wisconsin Mirror*, June 7, 1868; Dell Pilot, "Hops in the West," *Prairie Farmer*, n.s., 21, no. 9 (February 1868): 127; "Hop Culture," *Prairie Farmer*, n.s., 21, no. 18 (May 1868): 82.

48. F., "Hops," *Cultivator and Country Gentleman* 32, no. 816 (September 1868): 156–57. "F." was from Orleans County, New York.

49. "Hops in the Town of Fairfield in 1868," *Wisconsin Mirror*, May 26, 1869.

50. Hibberd, "Agriculture in Dane County," 154.

51. "High and Low," *Wisconsin Mirror*, September 23, 1868; "What Shall We Do?" *Wisconsin Mirror*, September 30, 1868; "About Future Hop-Growing," *Wisconsin Mirror*, October 7, 1868.

52. "Plow Them Up!" *Wisconsin Mirror*, November 18, 1868; "Plow Them Up," *Wisconsin Mirror*, December 23, 1868; "Hops in the West," *Prairie Farmer*, n.s., 21, no. 10 (March 1868): 152. Problems were also evident in Wisconsin in the 1869 harvest. See "Hops," *Cultivator and Country Gentleman* 33, no. 856 (June 1869): 475.

53. Samuel B. Ruggles, *Tabular Statements from 1840 to 1870, of the Agricultural Products of the States and Territories of the United States of America* (New York: Press of the Chamber of Commerce, 1874), 16. Many growers in New England and New York suffered from a decrease in prices in 1868 and 1869, and farmers in several sections plowed up their fields. "Hop Prospects," *Cultivator and Country Gentleman* 33, no. 848 (April 1869): 331; "Hops," *Cultivator and Country Gentleman* 33, no. 851 (May 1869): 375; "Hops," *Cultivator and Country Gentleman* 34, no. 870 (September 1869): 238.

54. A. R. Eastman, "A Paper on Hops, from Grubbing to Gathering," in *Fifteenth Annual Report of the New York State Agricultural Society, for the Year 1890* (Albany, N.Y.: James B. Lyon, 1891), 318–27.

55. Hubert Howe Bancroft, *The History of California* (San Francisco: History Company, 1884–90), vol. 6, 699. Although this claim has been generally accepted, it could be challenged. In December 1855 San Francisco legislators noted that "Mr. E. H. Taylor has a fine hop field under cultivation in this vicinity, or near Oakland. He is the first, we believe, to cultivate hops as a market crop in California." Dorothy H. Higgins, comp., "Continuation of the Annals of San Francisco, November 28, 1855, through December 31, 1855," *California Historical Society Quarterly* 17, no. 2 (June 1938): 176.

56. Daniel Flint wrote in later years, "I assure you that I had a very large field to work in and a

very small reservoir of information to draw from. I searched for all the written information I could procure. I interviewed every person I could find who had ever seen a 'hop-yard,' for I had never seen but two poles of hops in my life, and they were allowed each year to ripen and cure on the poles and to turn red and unsalable. The remembrance of those two poles that I saw in my boyhood days on the granite hills of New Hampshire, came very near to losing me my first crop of hops. I supposed that they had got to remain on the poles until they turned red before I could pick them, and if it had not been for a friend that came along, who was somewhat versed in hop culture . . ." *Pacific Rural Press* 21, no. 24 (June 1881): 416.

57. Daniel Flint, "The Hop Industry in California," *Transactions of the California State Agricultural Society During the Year 1891* (Sacramento: A. J. Johnston, 1892), 196. "The Farmer's Department," *Daily Appeal*, September 19, 1861, mentions that Daniel Flint's plants were from Vermont, and that Wilson Flint was responsible for bringing them to California in 1858 or 1857. See also W. A. T. Stratton, "The Pioneer Hop-Growers," *Pacific Rural Press*, September 5, 1891.

58. Daniel Flint, "Hop Growing in California," *Pacific Rural Press* 21, no. 24 (June 1881): 416; and "The Hop Industry in California," 196.

59. "California Hop Industry," *Pacific Rural Press* 42, no. 4 (July 25, 1891): 67; Otis Allen, "Early Hop Industry," *Pacific Rural Press* 42, no. 17 (October 1891): 346; *California Culturist* 1, no. 12 (November 1858): 273.

60. *Transactions of the California State Agricultural Society During the Year 1860* (Sacramento: Benjamin P. Avery, 1861), 239–40; Otis Allen, "Early Hop History," *Pacific Rural Press* 42, no. 17 (October 1891): 346; "The Sonoma Hop Growers," *Pacific Rural Press* 42, no. 7 (August 1891): 126. Allen and Bushnell challenged the Flints for the distinction of having the "pioneer hop yard," but the dispute was not easily settled. See Daniel Flint, "California Hop History: Who Is the Pioneer?" *Pacific Rural Press* 42, no. 8 (August 1891): 153.

61. "State Bounty Law," *Transactions of the California State Agricultural Society During the Year 1863* (Sacramento: O. M. Clayes, 1864), 58; *Transactions of the California State Agricultural Society During the Years 1864 and 1865* (Sacramento: O. M. Clayes, 1866), 404.

62. Flint, "Hop Growing in California," 416.

63. "Hop Circular," *Pacific Rural Press* 4, no. 7 (August 1872): 107.

64. L.H., "Hop Growing in California," *Cultivator and Country Gentleman* 36, no. 956 (May 1871): 307. The writer was contributing from San Francisco at the time.

65. Ibid.

66. In 1865 A. Cook planted the first hop vines in the county, at St. Helena, and in 1876 he sent a sample of his hops to the Centennial Exhibition in Philadelphia and received first prize. "Hops," *Napa Register*, July 1, 1876.

67. "Sacramento," *Pacific Rural Press* 42, no. 2 (July 1891): 30.

68. "California Hops," *Pacific Rural Press* 8, no. 15 (October 1874): 233; "Call from a Hop Grower," *Pacific Rural Press* 8, no. 16 (October 1874): 241.

69. "Restricting Hop Culture," *Pacific Rural Press* 10, no. 18 (October 1875): 280.

70. "Hops," *Pacific Rural Press* 16, no. 6 (August 1878): 88.

71. Peter J. Delay, *History of Yuba and Sutter Counties* (Los Angeles: Historical Record Company, 1924), 1187. See also the *Wheat Graphic* for March 24, April 21, and August 4, 1883; and "Death of Dr. D. P. Durst," *Daily Appeal*, June 12, 1910.

72. "Pleasonton *Times*, Sept. 15," *Pacific Rural Press* 56, no. 13 (September 1895): 189.

73. "The Largest Hop Crop per Acre," *Pacific Rural Press* 16, no. 6 (February 1891): 119.

74. U.S. Department of Agriculture, *Statistics*, Bulletin no. 225, 1959, tables 1–4.

75. Daniel Flint, "A Pest in the Hopyard," *Pacific Rural Press* 41, no. 6 (February 1891): 116.

76. Otis W. Freeman, "Hop Industry of the Pacific Coast States," *Economic Geography* 19, no. 3 (April 1936): 163. As will be seen in the last chapter, kilns from this period in California are extremely rare.

77. J. Neilson Barry, "Agriculture in the Oregon Country in 1795–1844," *Oregon Historical Quarterly* 30, (March–December 1929): 162; Charles Henry Carey, *History of Oregon* (Portland, Ore.: Pioneer Historical Publishing Company, 1922), 792.

78. "Hop Culture in Oregon," *Willamette Farmer* 2, no. 11 (May 1870): 83.

79. "First Hop Yard," *Eugene Weekly Guard*, September 9, 1899; Joseph Gaston, *The Centennial History of Oregon* (Chicago: S. J. Clarke, 1912), 1:544.

80. "One Pound of Hops," *Willamette Farmer*, November 18, 1871.

81. "The Hop Interest," *Willamette Farmer*, March 1873; "Hop Culture," *Oregon Cultivator* 3, no. 12 (May 1876): 3.

82. "Premiums on Hops at State Fair," *Oregon Cultivator* 3, no. 12 (May 1876): 3.

83. *Illustrated History of Lane County* (Portland, Ore.: A. G. Walling, 1884), 471.

84. Walker and Seaton, *Report on Agriculture*, 305–7.

85. "Alexander Seavey," in *Portrait and Biographical Record of the Willamette Valley, Oregon* (Chicago: Chapman, 1903), 908; "A Farewell to Hops," in *And When They're Gone* (Eugene, Ore.: Lane County Department of Employment and Training, 1980), 50.

86. Sidney W. Newton, *Early History of Independence, Oregon* (Salem, Ore.: Panther, 1971), 64; "Bird Island," *Independence Enterprise*, April 28, 1904. This area of Oregon was named by Nelson Jones because it was inhabited principally by "doves." The Horst ranch, north of the city, covered over 400 acres at the time.

87. "The Hop Industry," *Independence Enterprise*, July 23, 1903. The principal growers were H. Hirschberg, 100 acres; J. R. Cooper, 50 acres; Cooper Brothers, 40 acres; D. B. Taylor, 40 acres; and O. D. Rider, 40 acres; all were near Independence. See also "In the Hop Fields," *Northwest Pacific Farmer* 24, no. 27 (August 1894): 1.

88. U.S. Department of Agriculture, *Hops by States, 1915–56*, Statistical Bulletin no. 225 (Washington, D.C.: U.S. Department of Agriculture, 1958), 4–6; Paul Landis, "The Hop Industry: A Social and Economic Problem," *Economic Geography* 22, no. 2 (January 1939): 85.

89. Meeker, *Hop Culture in the United States*, 8; Frank L. Green, "Biographical Notes," in *Ezra Meeker, Pioneer: A Bibliographical Guide* (Tacoma: Washington State Historical Society, 1969), 11; "First Washington Hops," *Washington Standard*, September 23, 1887. Still another reference to Ezra Meeker's writing notes that it was "the spring of 1864": "Hop-Culture," *Washington Standard*, December 23, 1876.

90. "Hops," *Weekly Pacific Tribune*, October 9, 1874; Myrick, *The Hop*, 43; "Hops," *Washington Standard*, September 5, 1874; "A Puyallup Hop Farm," *Pacific Daily Tribune*, August 30, 1877; "Hop Fields of Puyallup and White River," *West Shore* 10, no. 11 (November 1884): 346.

91. J. P. Stewart was the third hop grower in the valley, having begun his culture in 1871; see Robert Montgomery, "History of the Puyallup," *Puyallup Valley Tribune*, July 30, 1904. The Klaber yards were the largest at the turn of the century, encompassing 130 acres. These were purchased and subdivided in 1906, forming the southern part of town, and mark the beginning of the end of the culture in western Washington. See "A Big Sale," *Puyallup Valley Tribune*, October 27, 1906.

92. "A Western Center," *West Shore* 16, no. 2

(February 1890): 1; *Washington Standard*, September 15, 1882; "Hop Picking," *Washington Standard*, September 24, 1886.

93. Carmen Olson, "Hop Growing Industry," *Issaquah Press*, November 11, 1954.

94. "Hops," *Washington Standard*, June 1, 1883; "The Largest Hop Crop per Acre," *Pacific Rural Press* 16, no. 6 (February 1891): 119.

95. Joyce Benjamin Kuhler, "A History of Agriculture in the Yakima Valley, Washington, from 1880 to 1900" (Master's thesis, University of Washington, 1940), 45–51.

96. Cindy Warren, "Charles Carpenter, An Early Hop Grower," in *People Who Have Made Yakima History*, ed. Gary L. Jackson (Yakima, Wash.: Yakima Herald Republic, 1978), 80–81.

97. *Washington Standard*, November 18, 1876; "Yakima and Its Surroundings," *West Shore* 13, no. 10 (October 1887): 715–27.

98. "A Great Hop Enterprise," *Pacific Rural Press* 41, no. 24 (June 1891): 54.

99. "Hop History," *North Pacific Rural Spirit*, July 18, 1902; "Hops in Washington," *Pacific Rural Press* 41, no. 11 (March 1891): 233; "In the Hop Fields," *Northwest Pacific Farmer* 23, no. 25 (August 1893): 8. In 1895 it was determined that the total area being cultivated by all the Meekers at Puyallup and on the White River was 611 acres. See "In the Hop Fields," *Northwest Pacific Farmer* 25, no. 11 (May 1895): 5.

100. The company incorporated with a capital stock of $120,000, and over $200,000 was spent in improvements. The officers included Richard Jeffs of White River, president; George W. Gove, vice-president and manager; and H. Dutard of San Francisco, treasurer. Others named were G. K. Baxter of Seattle and H. E. Levy of Victoria, British Columbia. "Hops in Washington," *Pacific Rural Press*, January 3, 1891.

101. "New Tacoma, Washington Territory," *Waterville Times*, December 13, 1877.

102. "The Hop Fly or Louse," *Pacific Rural Press* 41, no. 16 (April 1891): 376; "Pacific Coast Hop Notes," *Pacific Rural Press* 41, no. 21,

(May 1891): 502; "Washington Hop Notes," *Pacific Rural Press* 42, no. 6 (August 1891): 166.

103. "Hop Law," *Northwest Pacific Farmer* 29, no. 36 (September 1899): 14.

104. U.S. Department of Agriculture, *Statistics*, Bulletin no. 225, March 1958, tables 1–4.

105. Arnold and Penman, *History of the Brewing Industry*, 186–87.

106. Otis W. Freeman, "Hop Industry of the Pacific Coast States," *Economic Geography* 19, no. 5 (April 1936): 160.

107. Arthur L. Dahl, "Growing Hops in California: A Glimpse at an Industry That Will Be Hit by Prohibition," *Scientific American Supplement* 87, no. 2263 (May 1919): 312–13.

108. Landis, "The Hop Industry."

109. Hadley, "Hops in the United States," 46–49.

110. "Why Not a Beer with Hops," *Pacific Hop Grower* 3, no. 6 (October 1935): 2.

111. U.S. Department of Agriculture, *Statistics*, Bulletin no. 225, March 1958, tables 1–4.

112. Sam Churchill, "Valley Harvesting Sixty per Cent of Nation's Hops," *Yakima Herald*, August 30, 1964; Judy Ann Mills, "Valley May Grow All the Hops," *Yakima Herald Republic*, August 4, 1980; U.S. Bureau of the Census, *U.S. Census of Agriculture: 1959* (Washington, D.C.: U.S. Government Printing Office, 1960), 2:14.


## 2. Growing and Harvesting

1. F. W. Collins, "Hop Culture in the West," *Prairie Farmer*, n.s., 9, no. 8 (February 1867): 115; Dell Pilot, "Hop Culture," *Prairie Farmer*, n.s. 11, no. 18 (May 2, 1868): 82.

2. Warren Ferris, "Hop Growing," *Cultivator and Country Gentleman* 35, no. 902 (May 1870): 276–77.

3. The distance between rows seems to have been a matter of discussion from the beginning. See, for example, "Hops," *Genesee Farmer and Gardener's Journal* 2, no. 16 (April 1832): 122,

which recommends six to eight feet. Another article, "Culture of Hops," *Genesee Farmer and Gardener's Journal* 3, no. 7 (February 1833): 54, suggested that the "hills should be eight or nine feet asunder."

4. In the twentieth century there was considerably less variation wherever mechanical harvesting became common.

5. "Raising Hops from Seed," *Country Gentleman* 21, no. 25 (June 1863): 400; Walter Ferris, "Ferris' Seedling Hops," *Cultivator and Country Gentleman* 33, no. 836 (January 1869): 72–73. See also "Seedling Hop," *Cultivator and Country Gentleman* 28, no. 712 (September 1866): 153.

6. "Culture of Hops," *Genesee Farmer and Gardener's Journal* 3, no. 7 (February 1833): 54; F. B. Sinclair, "Planting Hop Roots," *Northwest Pacific Farmer* 23, no. 7 (April 1893): 1.

7. Abraham Hale Burgess, *Hops: Botany, Cultivation, and Utilization* (London: Leonard Hill; New York: Interscience Publishers, 1964), 19; Daniel Flint, "Hop Growing in California," *Pacific Rural Press* 21, no. 24 (June 1881): 416; L. R. Bayha, "Hop Culture," *Oregon Agriculturalist and Rural Northwest* 6, no. 13 (March 1897): 196.

8. F. W. Collins, "Culture of Hops," *Cultivator and Country Gentleman*, 31, no. 795 (April 1868): 263.

9. "The Picking of Hops," *Pacific Hop Grower* 3, no. 5 (September 1935): 4.

10. Dell Pilot, "Transplanting Hop Suckers," *Prairie Farmer* 20, no. 2 (July 1867): 18; Dell Pilot, "Hops in the West," *Prairie Farmer* 21, no. 9 (February 1868): 127.

11. For example, on the Horst ranch, near Independence, Oregon, 80 tons of fertilizer were shipped from San Francisco in April 1904, to be spread over 360 acres. Present-day growers use inorganic fertilizer.

12. "Hop Culture," *Genesee Farmer* 8, no. 3 (March 1847): 58; Daniel Flint, "Hop Growing in California," *Pacific Rural Press* 21, no. 24 (June 1881): 416.

13. H.C., "Hops," *Genesee Farmer and Gardener's Journal* 5, no. 33 (August 1835): 259; F.W.S., "About Hops," *Moore's Rural New-Yorker* 27, no. 16 (April 1873): 250; "Hops," *Wisconsin Mirror*, July 21, 1869; Daniel Flint, "Hop Culture in California," *Pacific Rural Press* 21, no. 24 (June 1881): 416.

14. Lincoln Cummings, "Hop Culture," *Genesee Farmer* 8, no. 4 (April 1847): 87. Cummings was writing from Augusta Township, in Oneida County, New York.

15. "Hop Culture in Central New York," *Waterville Times*, July 25, 1878.

16. "Culture of Hops," *Genesee Farmer and Gardener's Journal* 3, no. 7 (February 1833): 54; Warren Ferris, "Culture of Hops," *Cultivator and Country Gentleman* 35, no. 886 (January 1870): 21, and "Hop Culture in Central New-York," *Cultivator and Country Gentleman* 37 (September 1872): 627–28.

17. "Collins' Patent Horizontal Hop-Yard," advertisement appearing on the inside back cover of Ezra Meeker, *Hop Culture in the United States, Being a Practical Treatise on Hop Growing in Washington Territory* (Puyallup, Washington Territory: E. Meeker, 1883). It was first brought before the agricultural public in the *American Agriculturalist* in May 1864.

18. As early as 1831, Loudon mentioned that wires of iron or copper had been used in southern France; however, he did not think it was an improvement. See John Claudius Loudon, *An Encyclopedia of Agriculture: Comprising the Theory and Practice of the Valuation, Transfer, Laying Out, Improvement, and Management of Landed Property* (London: Longman, Rees, Orme, Brown, Green and Longman, 1831), 926.

19. F.W.C., "Hop Culture," *Cultivator and Country Gentleman* 36, no. 966 (July 1871): 469.

20. Ira C. Jenks, "Culture of Hops," *Cultivator and Country Gentleman* 31, no. 789 (April 1865): 321. Jenks stated that "with the horizontal string system, it is a fact that the vine will not run on the twine at all." See also Warren

Ferris, "Training the Hop," *Cultivator and Country Gentleman* 33, no. 851 (May 1869): 373. Ferris came to the conclusion based on his experiments that "the hop will not follow twine or anything else in a horizontal position."

21. F. W. Collins, "Hop Culture," *Prairie Farmer*, n.s., 19, no. 15 (April 1867): 235; F. W. Collins, "Culture of Hops," *Cultivator and Country Gentleman* 31, no. 795 (April 1868): 263; E. A. Avery, "Training Hops Horizontally," *Cultivator and Country Gentleman* 29, no. 750 (May 1867): 347; "Foreign Notices," *Cultivator and Country Gentleman* 39, no. 731 (January 1867): 49.

22. L. D. Snook, "Snook's System of Hop-Training," *Cultivator and Country Gentleman* 31, no. 785 (January 1868): 74. Alterations to this system were proposed by J. W. Clarke, "Hop-Poles: Simple Plan for Training Hop-vines," *Cultivator and Country Gentleman* 31, no. 792 (March 1868): 202–3.

23. C.D.P., "Training Hops," *Cultivator and Country Gentleman* 31, no. 789 (February 1868): 146.

24. "Prospects of the Crop," *Wisconsin Mirror*, September 9, 1868.

25. L.H., "Hop Growing in California," *Cultivator and Country Gentleman* 36, no. 956 (May 1871): 307.

26. Daniel Flint, "The Hop Industry in California," *Transactions of the California State Agricultural Society During the Year 1891* (Sacramento: A. J. Johnston, 1892): 197.

27. "The State Hop Crop," *Washington Standard*, September 29, 1893.

28. "Hop Growers' Pointers," *Northwest Pacific Farmer* 29, no. 36 (September 1899): 14; L. R. Bayha, "Hop Culture," *Oregon Agriculturalist and Rural Northwest* 6, no. 15 (April 1897): 219; "A Bulletin on Hops," *Oregon Agriculturalist and Rural Northwest* 7, no. 15 (April 1898): 227.

29. "Hop District," *West Side Enterprise*, February 18, 1904.

30. The process of harvesting will be dealt with at greater length in the succeeding chapters.

31. "Among the Hop Fields," *Waterville Times*, August 23, 1878.

32. F. M. Blodgett, *Hop Mildew*. Cornell University Agricultural Experiment Station Bulletin no. 328 (Ithaca, N.Y.: Cornell University, 1913), 281–82.

33. L. D. Wood, *Hops: Their Production and Uses* (Chicago: National Hops Company, 1938), 29–30.

34. Z. E. Jameson, "Hop-Growing," *Cultivator and Country Gentleman* 29, no. 733 (January 1867): 75; "Hop Growers' Pointers," *Northwest Pacific Farmer* 29, no. 36 (September 1899): 14. Whale oil and quassia chips were the most common in the Northwest, if the records of Ezra Meeker are any indication.

35. James Whorton, *Before Silent Spring: Pesticides and Public Health in Pre-DDT America* (Princeton, N.J.: Princeton University Press, 1974), 24. See also: Thomas R. Dunlap, *DDT: Scientists, Citizens, and Public Policy* (Princeton, N.J.: Princeton University Press, 1981). The author extends his thanks to Douglas Hunt for these references.

36. Daniel Flint, "Hop Culture in California," *Pacific Rural Press* 21, no. 24 (June 1881): 416; "Hop Crop," *Pacific Rural Press* 56, no. 11 (September 1895): 170.

37. "Hop Culture on the Puyallup," *Washington Standard*, July 8, 1876.

38. "Hop Picking," *Washington Standard*, September 24, 1886.

39. "Culture of Hops," *Genesee Farmer and Gardener's Journal* 3, no. 7 (February 1833): 54. This is copied in *Cultivator* 10, no. 3 (March 1843): 44, with little alteration.

40. *Country Gentleman* 12, no. 9 (September 1858): 138–39.

41. Edward J. Lance, *The Hop Farmer; or, A Complete Account of Hop Culture, Embracing Its History, Laws, and Uses: A Theoretical and*

*Practical Inquiry into an Improved Method of Culture, Founded on Scientific Principles* (London: Joseph Rogerson, 1838), 114.

42. "Regulating the Size of Hop Boxes," *Waterville Times*, March 30, 1876.

43. "Size of Hop-Boxes," *Wisconsin Mirror*, August 5, 1868.

44. "Juneau Co. Hop Growers' Association," *Wisconsin Mirror*, September 2, 1868.

45. "Hints on Hop-Growing, no. 6," *Pacific Rural Press* 8, no. 2 (July 1874): 25.

46. "The Hop Yard," *Oregon Agriculturalist and Rural Northwest* 11, no. 24 (September 1902): 382.

47. Meeker, *Hop Culture in the United States*, 19.

48. "A Handy Hop Bag," *Puyallup Valley Tribune*, September 24, 1904. The nature of picking containers is considered at length in chap. 4.

49. Karl G. Becke, "History and Present Status of the Oregon Hop Industry" (Senior thesis, University of Oregon, 1917), 5.

50. Daniel Flint, "Hop Culture in California," *Pacific Rural Press* 21, no. 24 (June 1881): 416. The role of mechanical pickers will be examined in chap. 3. The vines, having had their fruit stripped, were often left on the poles to dry and later were cut off and chopped or burned, the ashes to be applied to the fields.

51. See for example, Stephen Powers, "On the Cultivation of Hops," *Genesee Farmer* 19, no. 7 (July 1858): 919. An early reference to the disadvantages associated with natural drying suggests that kilns were preferred in New York State almost from the establishment of the industry. "Hops that are dried in the sun, lose their rich flavor, and if under cover, they are apt to ferment and change with the weather and lose their strength." "Gathering and Curing Hops," *Cultivator* 1, no. 1 (March 1834): 10.

52. The development of the frame kiln was related to improvements in heating technology, as will be seen in chap. 5.

53. As was often the case with hop presses, hop stoves were in such high demand in the various growing regions that small businesses often arose to supply them. In Woodburn, Oregon, for example, John McKinney began manufacturing the Eclipse Hop Stove in 1894 and continued for over a decade, expanding his foundry and machine works to accommodate the business. "The Eclipse Hop Stove," *Oregon Farmer* 23, no. 20 (July 1914): 9.

54. Otis W. Freeman, "Hop Industry of the Pacific Coast States," *Economic Geography* 19, no. 5 (April 1936): 158.

55. "Hops," *Pacific Rural Press* 21, no. 25 (June 1881): 434.

56. "Price for Drying Hops," *Wisconsin Mirror*, August 5, 1868.

57. "Hop Growers' Pointers," *Northwest Pacific Farmer* 29, no. 36 (September 1899): 14.

58. Ibid.

59. Otis Freeman, "Hop Industry of the Pacific Coast States," *Economic Geography* 19, no. 5 (April 1936): 158.

60. Sulphur first came into widespread use in the late 1850s as a powdered disinfectant to prevent mold and destroy insects, and it was employed for this purpose well into the twentieth century throughout Europe and the United States. Its use in the drying process was primarily to bleach the hops, rendering an inferior crop equal or similar in appearance to a first-class product. While growers and dealers were interested in obtaining the best price for their hops, brewers generally objected to the use of sulphur on the grounds that it changed the physical and chemical properties of the hops. Perhaps most obvious, it was claimed that sulphurous acid killed the yeast in the fermentation process and thus slowed production. One solution for this, proposed by W. A. Lawrence, was to make more extensive use of hop extracts in brewing. Lawrence's New York Extract Company will be discussed further below. Another alternative, first proposed and promoted by

Courtland L. Terry of Waterville, was the "cold process" of bleaching hops. Terry demonstrated that, by simply using a small kerosene stove to burn sulphur in a common kitchen dripping pan *without* heating the kiln, and keeping all the vents of the kiln closed, be could bleach his hops more completely and quickly than by using the standard method. "Sulphured Hops," *Waterville Times*, March 23, 1876; "Among the Hop Fields," *Waterville Times*, August 22, 1878; "The Hop Crop," *Freeman's Journal*, September 5, 1878. See also "Picking and Drying Hops," *Wisconsin Mirror*, August 5, 1868; and "Hop Growers' Pointers," *Northwest Pacific Farmer* 29, no. 36 (September 1899): 14, for information regarding developments outside of New York.

61. "Picking and Drying Hops," *Wisconsin Mirror*, August 5, 1868.

62. "Culture of Hops, no. 3," *Country Gentleman* 1, no. 10 (March 1853): 146–47.

63. "On the Raising of Hops," *Transactions of the Society for the Promotion of Agriculture, Arts and Manufactures* (Albany, N.Y.: Charles R. and George Webster, 1801), 1:149–50.

64. The same material was repeated without amendment in the *Cultivator* 1, no. 1 (March 1834): 2; and with only minor amendments in the *Cultivator* 2, no. 12 (February 1836): 190.

65. *Eighty Years' Progress of the United States: A Family Record of American Industry, Energy, and Enterprise* (Hartford, Conn.: L. Stebbins, 1868), 1:88.

66. H.C., "Hops," *Genesee Farmer and Gardener's Journal* 5, no. 33 (August 1835): 259.

67. Lincoln Cummings, "Hop Culture," *Genesee Farmer* 8, no. 7 (July 1847): 158.

68. In 1853 this procedure was followed in Otsego County: "The hops are let down into a strong curb or box of the required size, (generally about 4½ feet in length, 4 feet in height, and 17 inches in width,) a piece of baling cloth first being laid on the bottom of the curb, of the right length for the bale; two men enter the curb and tread the hops as they are let down, until it is full. Another piece of cloth is then laid on the top of the hops, a follower put on and pressed down with a screw, (fixed for the purpose in a press similar to a cider press,) till the two edges of the cloth will meet. The curb (being constructed for the purpose,) is then taken apart, the screw holding the hops to their place; the edges of the cloth are lapped together, and secured completely round the bale. After it is sewed, the screw is taken up, the bale taken out, the ends cased, and it is finished." (An Otsego Hop Grower, "Culture of Hops, no. 3," *Country Gentleman* 1, no. 13 [March 1853]: 194.) In 1861 a description reads: "A press about five feet long and four feet deep, by eighteen inches, is necessary. Any carpenter can make one, and any screw will answer, of which there is more or less in every neighborhood. The press should be made so as to take down the sides, then take some sacking and spread on the bottom, letting it come beyond the press about eighteen inches on each side, then set up the press, fill with hops and press down; fill again and press, and on until the bale is about three feet high. The last time it is filled put on a piece of sacking, the same as the bottom, and after it is pressed down, take down the sides and bring the sacking together, and sew with twine; then roll it out and sew up the ends, and they are ready for market." ("Curing Hops," *Working Farmer* 13, no. 9 [September 1861]: 204.)

69. "Call From a Hop Grower," *Pacific Rural Press* 8, no. 16 (October 1874): 241; George Clinch, *English Hops: A History of Cultivation and Preparation for the Market from the Earliest Times* (London: McCorquodale, 1919), 38.

70. "Hops," *Waterville Times*, May 21, 1859.

71. "Harris' Hop Press," *Waterville Times*, August 25, 1860.

72. Messrs. Buckingham, Bright and Company, of Waterville were noted for manufacturing the Harris device. (*Waterville Times*, August 10, 1865.) In Wisconsin the "celebrated" Harris Hop Press was manufactured by C. J.

Hanson of Kilbourn City as early as July, 1868. See Advertisement, *Wisconsin Mirror*, July 22, 1868. The Morrison press seems to have been a favored alternative somewhat later in the Pacific Northwest. ("In the Hop Fields," *Northwest Pacific Farmer* 24, no. 13 [May 1894]: 5.) The "M & M," an improved Morrison press, was manufactured in McMinnville, Oregon ("The Hop Yard," *Oregon Agriculturalist and Rural Northwest* 6, no. 21 [July 1897]: 323; *Independence Enterprise*, September 12, 1895).

73. Daniel Flint, "Hop Culture in California," *Pacific Rural Press* 21, no. 24 (June 1881): 416. Flint's description of his rack-and-pinion press is instructive: "It stands upright, flush with the floor, and as soon as the horses can walk 100 ft. with the end of the rope the follower is down, the doors thrown open and ready for sewing." Two horse-powered presses were constructed in the Russian River valley in 1881 ("Hops," *Pacific Rural Press* 21, no. 25 [June 1881]: 434). For Beardsley's earlier steam-powered press, see *Waterville Times*, March 26, 1868.

74. "Hops in Washington," *Pacific Rural Press* 41, no. 11 (March 1891): 233; Herbert Myrick, *The Hop: Its Culture and Cure, Marketing and Manufacture* (New York: Orange Judd, 1914), 206.

75. "The Management of Hops," *Working Farmer* 2, no. 11 (January 1851): 251. The idea surfaced again; see "Hops: Preparation for Market," *Working Farmer* 7, no. 2 (March 1855): 3; "Hops," *Working Farmer* 12, no. 11 (November 1860): 253.

76. *Cultivator* 2, no. 6 (August 1835): 89–90.

77. "Hops: How They May Be Preserved," *Annual Report of the American Institute, of the City of New York for the Years 1865 '66* (Albany, N.Y.: C. Wendell, 1866), 127–29; U.S. Patent Office, *Subject Index of Patents for Inventions, 1790–1873* (New York: Arno Press, 1976), 2:738.

78. "Local Brevities," *Waterville Times*, October 5, 1876.

79. Ibid., May 31, 1877 and February 21, 1878.

80. F. W. Collins, "Hops: Statistics of Culture," *Annual Report of the American Institute, of the City of New York, for the Year 1866–67* (Albany, N.Y.: Van Benthuysen and Sons, 1867), 173; W. A. Lawrence, "The Preservation of the Hop in Waterville, New York," in Meeker, *Hop Culture in the United States*, 111–14.

81. "History of Hop Extract Plant in New York," *Pacific Hop Grower* 4, no. 6 (October 1936): 2. This article was copied from the *Waterville Times and Hop Reporter*. Other patents both in the United States and abroad took a slightly different approach. Charles A. Seeley, a New York chemist, and W. S. Newton, an English experimenter, claimed to have invented an alternative method of obtaining the extract of hops by the discovery that light products of petroleum were rapid and complete solvents of the essential oils. Hops steeped in naphtha under moderate heat—less than 120 degrees Fahrenheit—gave up their special chemical compounds. There was no known commercial exploitation of this idea, however, perhaps because of the complications these processes introduced. See: "New Mode of Obtaining the Extract of Hops," *Pacific Rural Press* 2, no. 13 (September 1871): 194; "Seeley's Extract of Hops," *Pacific Rural Press* 4, no. 6 (August 1872): 83.

*3. The Grower's Perspective*

1. James H. Dunbar, "Culture of the Hop," *Cultivator* 8, no. 5 (May 1841): 84. This is one of the most thorough early financial accounts that have come to light.

2. "Premium Farm: Hops, &c," *Country Gentleman* 2, no. 27 (July 1853): 6. The same information was carried in "Profitable Crop of Hops," *Cultivator* 1, no. 8 (August 1853): 238.

3. Warren Ferris, "Hop Growing," *Country Gentleman* 35, no. 898 (April 1870): 211.

4. "In the Hop Fields," *Northwest Pacific Farmer* 25, no. 11 (May 1895): 5.

5. "Little Response to Organization Plea," *Pacific Hop Grower* 4, no. 11 (April 1937): 2. The travel of insurance men on the road from Salem to Buena Vista, Oregon, made the road so dusty that the regional newspaper reported half-jokingly that the residents on either side were thinking of getting a road sprinkler and asking the agents to help pay for it. "In the Hop Fields," *Northwest Pacific Farmer* 24, no. 27 (August 1894): 1.

6. Judge S. Cheever, "Hop Culure," *Cultivator* 4, no. 2 (April 1837): 39. This was copied under the same title by *Genesee Farmer* 7, no. 15 (April 1837): 120.

7. H.C., "Hops," *Genesee Farmer and Gardener's Journal* 5, no. 33 (August 1835): 259.

8. U.S. Department of Agriculture, *Hops by States, 1915–56* (Washington, D.C.: U.S. Department of Agriculture, 1958), 5.

9. "The Hop Market," *Vermont Family Visitor* 1, no. 8 (January 1846), 226.

10. See, for example, G. W. Ryckman's advertisement in the *Madison Observer*, September 7, 1848, 3, and in the *Freeman's Journal*, September 16, 1848, 3. Located in New York, he was selling hops, barley, and malt to English, French, and German brewers.

11. The William P. Locke Papers are on deposit in the Archives of Colgate University, Hamilton, New York.

12. "Millions in Farming," *Waterville Times*, September 9, 1873, 2.

13. "Hop Buyers," *Wisconsin Mirror* 2, no. 9 (September 1869), 3. Weaver, in turn, went to the Pacific slope in 1879 to visit the hop-growing regions of California and Oregon. There he purchased about fifteen carloads and shipped them to the breweries at Milwaukee. "Hon. Richard Weaver," in *Portrait and Biographical Record of Waukesha County, Wisconsin* (Chicago: Excelsior, 1894), 200.

14. "A Good Move," *Wisconsin Mirror*, 1, no. 12 (September 16, 1868), 3. See, for example, the advertisements of Charles Ehlermann and Company of St. Louis, Missouri, and the Dole Brothers of Boston, Massachusetts, in the *Freeman's Journal*, August 28, 1879, 1.

15. Ezra Meeker Papers, Box 3, Folder 23, Telegrams, Washington State Historical Society, Tacoma.

16. "Millions in Farming," *Waterville Times*, September 9, 1875, 2; "Hops," *Waterville Times*, August 1, 1878, 2.

17. "Hops," *Freeman's Journal*, September 18, 1879, 3; *Freeman's Journal*, September 23, 1882, 3.

18. Daniel Flint, "Hop Growing in California," *Pacific Rural Press* 21, no. 24 (June 1881): 416.

19. "Hop-Culture on the Puyallup," *Washington Standard*, July 8, 1876, 2; "Hops," *Daily Pacific Tribune*, July 11, 1876, 2.

20. "In the Hop Fields," *Northwest Pacific Farmer* 24, no. 24 (August 1894): 1. Less well known was the *Hop-Growers' Journal*, edited for a time by G. H. Andrews.

21. "The Hop Yard," *Oregon Agriculturalist and Rural Northwest* 7, no. 23 (August 1898), 3; "Drought Will Cut Oregon Hop Crop," *Oregonian*, August 13, 1906, 1.

22. Warren Ferris, "Sharp Practice in Buying Hops," *Cultivator and Country Gentleman* 39, no. 1133 (October 1874): 644.

23. "In the Hop Fields," *Northwest Pacific Farmer* 30, no. 3 (January 1900): 9.

24. "Hops," *Cultivator and Country Gentleman* 34, no. 867 (September 1869): 178.

25. "Important to Hop Growers!" *Waterville Times*, January 17, 1879, 2.

26. Ibid. On the other hand, some farmers also used means that were less than perfectly honorable to better their position. Perhaps the most obvious was the abuse of the "sample system." A grower with ten bales would give samples to say, six different dealers, each of whom would send them to New York, where they appeared to represent sixty bales. In effect, these six samples were bidding against one another, overinflating the market but guaranteeing the grower the best possible price. "Abuse of the

Sample System," *Oregon Agriculturalist and Rural Northwest* 7, no. 6 (December 1897), 91.

27. Ninetta Eames, "The Hop Industry in Mendocino," *Pacific Rural Press* 42, no. 24 (December 1891): 490.

28. "Hops in Washington," *Pacific Rural Press* 41, no. 11 (March 1891): 234.

29. "Hop Contracts Void," *Independence Enterprise*, October 30, 1902, 4. The terms cited here were common in many contracts of the period.

30. "A Word with Our Hop-Growers," *Pacific Rural Press* 42, no. 1 (July 1891): 3.

31. "Hop Law," *Northwest Pacific Farmer* 29, no. 36 (September 1899): 14.

32. "Hop Growers' Association," *Waterville Times*, June 4, 1874, 2; "Much Needed," *Waterville Times*, September 10, 1874, 3; "Hop Growers' Convention!" *Waterville Times*, October 14, 1875, 3; "Where the Hops Are Raised," *Waterville Times*, April 20, 1876, 3. The second annual meeting of the "Hop Growers' Union of Central New York" was noted in the *Waterville Times*, January 11, 1877, 3; however, attendance was small.

33. "Hops," *Freeman's Journal*, September 11, 1879, 3.

34. "The Hop Yard," *Oregon Agriculturalist and Rural Northwest*, August 15, 1901, 366, notes the twenty-fourth annual Central New York Hop Grower's Association meeting at Sylvan Beach, with an estimated attendance of 18,000.

35. "Hop Growers' Club," *Wisconsin Mirror* 1, no. 4 (July 22, 1868): 3; "Meeting of Hop Growers, Juneau Co. Hop Growers' Association," *Wisconsin Mirror* 1, no. 10 (September 1868): 3.

36. "Hop Growers' Meeting," *Pacific Rural Press* 42, no. 3 (July 1891): 54. Two years later, dissatisfied with the prices offered by San Francisco dealers, Mendocino growers chartered a vessel to load hops for England. "Direct Shipment," *Northwest Pacific Farmer* 23, no. 30 (October 1893): 5.

37. "The Sonoma Hop Growers," *Pacific Rural Press*, August 15, 1891, 126.

38. "Hop Growers Will Organize," *Northwest Pacific Farmer* 23, no. 27 (August 1893): 8.

39. "Money in Hops," *Oregonian*, July 27, 1893, 3; "Splendid Outlook," *Oregonian*, April 14, 1894, 2; "In the Hop Fields," *Pacific Farmer* 24, no. 25 (August 1894): 1.

40. "Hop Grower's Meeting," *Independence Enterprise*, May 9, 1895, 3; "Hop-Grower's Association," *Independence Enterprise*, July 11, 1895, 3.

41. "Hop Growers," *Northwest Pacific Farmer* 29, no. 39 (November 1899): 1, and "Oregon Hop Growers Organize," *Oregon Agriculturalist and Rural Northwest* 9, no. 4 (November 1899): 62. The main office of the organization was moved to Portland in early 1900; "In the Hop Fields," *Northwest Pacific Farmer* 31, no. 62 (October 1901): 4.

42. "The Hop Yard," *Oregon Agriculturalist and Rural Northwest* 9, no. 13 (March 1900): 226. The reason for dropping the plan was reportedly the spread of blue mold among the cured hops. Joseph A. Hill, "The Oregon Hop Industry," *Scientific American Supplement* 51, no. 1317 (March 30, 1901), 21113; "The Hop Situation," *Puyallup Valley Tribune*, July 29, 1905, 1.

43. "Important to Hop Growers," *Independence Enterprise*, October 16, 1902, 4.

44. "A Big Hop Contract," *Independence Enterprise*, April 14, 1904, 3.

45. Karl G. Becke, "History and Present Status of the Oregon Hop Industry" (Senior thesis, University of Oregon, 1917).

46. "Puyallup Hops and Pickers," *Daily Pacific Tribune*, August 20, 1877, 3.

47. "A Puyallup Hop Farm," *Daily Pacific Tribune*, August 30, 1877, 2.

48. "In the Hop Fields," *Northwest Pacific Farmer* 24, no. 15 (June 1894): 1.

49. "The Hop Yard," *Oregon Agriculturalist and Rural Northwest* 9, no. 11 (February 1900): 174.

50. G. R. Hoerner, "Hop Growers Organizations," *Pacific Hop Grower* 4, no. 1 (May 1936): 3, 5; "To Unite or Not to Unite," *Pacific Hop Grower* 3, no. 8 (December 1935): 2.

51. "New York Legislature Passes Hop Bill," *Pacific Hop Grower* 3, no. 11 (March 1936): 2.

52. "New York Growers Form Cooperative," *Pacific Hop Grower* 5, no. 1 (June 1937): 2; "New York's Hop Industry Is Forging Ahead," *Pacific Hop Grower* 5, no. 3 (August 1937): 5.

53. "It's Up to the Grower Now," *Pacific Hop Grower* 5, no. 11 (April 1938): 2.

54. "Agreement Tried," *Pacific Hop Grower* 6, no. 6 (November 1938): 4.

55. Anglo-American Council on Productivity, *The Hop Industry* (London and New York: Productivity Team, 1951), 80.

56. "Destructive Fire," *Waterville Times*, October 3, 1857, 3. The article, one of several of its kind, describes the destruction of the Cornelius Carter residence, in the town of Sangerfield, and in so doing mentions the number of pickers and the loss of their belongings.

57. Mary E. Tooke, Oneida Castle, to Manley J. Tooke, Sheridan, N.Y., September 3, 1882, in the Tooke Family Papers, Department of Manuscripts and Archives, John M. Olin Library, Cornell University.

58. Obituary, James F. Clark, *Freeman's Journal*, April 19, 1900, 3.

59. "James F. Clark, Mayor of Hop City, The Largest Single Hop Grower in the Country," *Grip's Valley Gazette* 1, no. 2 (October 1893): 2–5.

60. James Fenimore Cooper, *Reminiscences of Mid-Victorian Cooperstown* (Cooperstown, N.Y.: Otsego County Historical Society, 1936), 36–37.

61. "The Tramp Nuisance: Public Meeting," *Freeman's Journal*, August 29, 1878, 3.

62. Ibid.

63. "Buy a Revolver and Shoot the Tramps," *Freeman's Journal*, September 19, 1878, 3.

64. "The Hop Harvest: Serious Problems for Growers," *Freeman's Journal*, September 10, 1881, 3.

65. "The Tramp Nuisance," *Freeman's Journal*, August 22, 1901, 3; "Hop-Pickers: What They Are, and How They Conduct Themselves," *Waterville Times*, September 19, 1878, 2. The article was originally written for the *Utica Herald*.

66. Ibid.

67. "In the Hop Fields," *Northwest Pacific Farmer* 23, no. 27 (August 1893): 8.

68. "Picking Will Begin Soon," *Puyallup Valley Tribune*, September 3, 1904.

69. "Lack of Pickers May Ruin Many Hops," *Pacific Hop Grower* 3, no. 5 (September 1935): 6; "CCC Hop Picking Scheme Doubtful," *Pacific Hop Grower* 5, no. 4 (September 1937): 3.

70. *Puyallup Valley Tribune*, September 19, 1903; *Independence Enterprise*, September 12, 1895.

71. "Hop-Pickers: What They Are, and How They Conduct Themselves," *Waterville Times*, September 19, 1878.

72. "Hop Crop in This Region," *Wisconsin Mirror*, August 28, 1868.

73. "Villainous," *Waterville Times*, August 8, 1862.

74. "Disastrous Fire," *Waterville Times*, September 12, 1874; "Another Fire," *Waterville Times*, September 26, 1876.

75. Ninetta Eames, "The Hop Industry in Mendocino County," *Pacific Rural Press* 42, no. 24 (December 1891): 490.

76. F. M. Dupont, "Hops: Their Chemical Makeup and Brewing Value," *Pacific Hop Grower* 3, no. 7 (November 1935): 3.

77. "The Hop Crop," *Freeman's Journal*, September 5, 1878.

78. "Trial of Hop-Pickers," *Wisconsin Mirror*, July 29, 1868; "Hop-Picking Machines," *Wisconsin Mirror*, September 9, 1868. Yet another machine was described in "New Hop Picker," *Wisconsin Mirror*, August 12, 1868; it was the invention of Amyntus Briggs of New Buffalo, Sauk County.

79. *Oneida Dispatch*, June 7, 1878.

80. *Oneida Dispatch*, August 2, 1878. "Locke's Hop Picking Machine," *Waterville*

*Times*, July 25, 1878, noted that this machine consisted of two rubber rollers, so constructed as to draw in the branches while two steel rollers having the opposite action picked the hops.

81. "Machine-picked vs. Hand-picked Hops," *Scientific American Supplement* 68, no. 1770 (December 1909): 361.

82. "Experimental Machine Picks and Dries Hops," *Pacific Hop Grower* 5, no. 6 (November 1937): 4.

83. Neil Joyce, "Hop Ranching in Mendocino County Is Large Business," *Redwood Journal-Press Dispatch*, October 19, 1950.

84. "Oregon Hop Men View Robot Picker," *Pacific Hop Grower* 6, no. 6 (November 1938): 3.

85. Anglo-American Council on Productivity, *The Hop Industry*, 9. Sam Churchill, "Valley Starts Record Hop Crop Harvest," *Yakima Sunday Herald*, August 24, 1958.

## 4. The Pickers

1. "Health of Hop Growers," *New England Farmer and Gardener's Journal* 24, no. 41 (April 1836): 325.

2. In England, adults who took on various tasks in the hop harvest made it their business to specialize in a limited part of the process. Picking was labor intensive, but there were plenty of hands available in London and its suburbs.

3. [John J. Skinner,] "The Hop," *American Farmer* 2, no. 39 (December 1820): 305.

4. Howard S. Russell, *A Long, Deep Furrow: Three Centuries of Farming in New England* (Hanover, N.H.: University Press of New England, 1976), 202, 226, 314, 406.

5. "Hops and Their Culture," *Prairie Farmer*, n.s., 17, no. 10 (March 1866): 145. Commenting on the hop fields of Illinois, the writer noted that "boys are too boisterous generally, and not counted profitable hands."

6. "The Culture of the Hop, no. 2," *Country Gentleman* 1, no. 7 (February 1853): 99. This was also included, under the same title, in *Cultivator* 1, no. 3 (March 1858): 82–83. Although the agricultural press of the period was dominated by men, the pleasurable aspects of picking have been documented from the women's point of view in the twentieth century in oral interviews.

7. "Hops in America," *Working Farmer* 15, no. 3 [whole no. 171] (March 1863): 59.

8. *Collections of the History of Albany, from Its Discovery to the Present Time* (Albany, N.Y.: J. Munsell, 1865), 1:488. This quotes the *Albany Evening Journal*, August 28, 1860, which noted the "demand for a large number of females to go to the country and assist in gathering hop crops. The intelligence offices in this city have furnished in the neighborhood of four hundred women and girls, who have been taken to Otsego, Herkimer, Oneida and Madison Counties in this state, while some have been taken to the Eastern states."

9. John Rooney, "Hop Industry in Detail," in *A Standard History of Sauk County, Wisconsin*, ed. Harry Ellsworth Cole (Chicago and New York: Lewis, 1918), 98.

10. "Among the Hop Fields," *Waterville Times*, August 23, 1878.

11. John Rooney, "Hop Industry in Detail," 97.

12. James H. Smith, *History of Chenango and Madison Counties, New York* (Syracuse, N.Y.: D. Mason, 1880), 95.

13. "Hop Pickers," *Wisconsin Mirror*, August 26, 1868.

14. "Hop-Pick'n-Brokers," *Independent*, August 6, 1867.

15. The term *city pickers* was the most common, however the terms *foreign pickers* or *off pickers* were also used in the East.

16. Helen Butler, "Hop Picking on the James Tooke Farm," ca. 1963 Department of Manuscripts and Archives, Cornell University, Ithaca, N.Y., [6].

17. "Hop Pickers," *Wisconsin Mirror*, August 5, 1868; "Hops-Pick'n-Brokers,"

*Independent,* August 6, 1867; "30,000 Rallying to the Poles!!" *Wisconsin Mirror,* September 2, 1868.

18. "Trouble Between Hop Growers and Pickers!" *Wisconsin Mirror,* September 13, 1868.

19. "James F. Clark, Mayor of Hop City, The Largest Single Hop Grower in the Country," *Grip's Valley Gazette* 1, no. 2 (October 1893): 2–5.

20. "Hop-Picking," *Independent,* September 3, 1867.

21. Ninetta Eames, "The Hop Industry in Mendocino," *Pacific Rural Press* 42, no. 24 (December 1891): 490.

22. Ninetta Eames, "In Hop-Picking Time," *Cosmopolitan* 16, no. 1 (November 1893): 30. Growers in the Russian River valley were less involved with Oriental pickers.

23. Reuben Ibanez, ed., *Historical Bok Kai Temple in Old Marysville, California* (Marysville, Calif.: Marysville Chinese Community, 1967). Not only the Bok Kai Temple but also several houses in Marysville were built in a traditional Chinese manner.

24. Daniel Flint, "Hop Growing in California," *Pacific Rural Press* 21, no. 24 (June 1881): 416; "China vs. Indian Hop Picking," *Pacific Rural Press* 10, no. 11 (September 1875): 165.

25. Eames, "Hop Industry," *Pacific Rural Press* 42, no. 24 (December 1891): 490.

26. "The Chinamen," *Napa Register,* March 25, 1876.

27. "How Not to Do It," *Napa Register,* April 29, 1876.

28. "Chinese Cheap Labor," *Anaheim Daily Gazette,* August 10, 1877; "Conflict of Races," *Anaheim Daily Gazette,* August 19, 1877.

29. Advertisement, *Daily Appeal,* July 25, 1889.

30. "The Hop Yard," *Oregon Agriculturalist and Rural Northwest,* September 1, 1907: 382.

31. "The Sonoma Hop Growers," *Pacific Rural Press* 42, no. 7 (August 1891): 126.

32. Daniel Flint, "The Hop Industry in California," in *Transactions of the California State Agricultural Society During the Year 1891* (Sacramento: A. J. Johnston, 1892), 197. The number of Asian immigrants in the fields was always difficult to determine and the public's perception was often clouded. See "Pacific Coast Hop Notes," *Pacific Rural Press* 41, no. 21 (May 1891): 502, quoting the *Sacramento Bee.*

33. "In the Hop Yards," *Pacific Rural Press* 56, no. 6 (August 1898): 87.

34. Eames, "In Hop-Picking Time," 27.

35. "Hop Pickers Wanted," *Pacific Rural Press* 56, no. 6 (August 1898): 87.

36. "Annual Arrival," *West Side Enterprise,* September 1, 1904.

37. "Moving Hop Pickers," *Eugene Weekly Guard,* September 9, 1899.

38. "A Farewell to Hops," in *And When They're Gone* (Lane County, Ore.: Land County Department of Employment and Training, 1980), 46–60.

39. "All Quiet in Butteville," *Oregonian,* September 6, 1893; "Chinese Hop Pickers," *Oregonian,* September 8, 1893.

40. Annie Marion MacLean, "With Oregon Hop Pickers," *American Journal of Sociology* 15 (July 1909): 84.

41. Ibid.

42. Ibid., 94.

43. Early references to the introduction of Chinese pickers in the Puyallup Valley make it clear that native Americans were preferred. See "The Puyallup Hop Crop," *Daily Pacific Tribune,* August 30, 1875; "Hops," *Daily Pacific Tribune,* September 6, 1875; *Washington Standard,* September 9, 1876.

44. *Washington Standard,* August 25, 1877 and September 15, 1877; "Sumner Guards," *Washington Standard,* August 17, 1878.

45. "Hop Culture on the Pacific Coast," *West Shore* 14, no. 9 (September 1888): 466–71; "Picturesque Hop Pickers," *Puyallup Valley Tribune,* September 10, 1904.

46. W. H. Bull, "Indian Hop-Pickers on Puget Sound," *Harper's Weekly* 36, no. 1850 (July 1892): 545–46.

47. *Puyallup Valley Tribune*, September 17, 1904.

48. *Pacific Rural Press* 16, no. 1 (January 1891): 2.

49. Susan Lord Currier, "Some Aspects of Washington Hop-Fields," *Overland Monthly* 32 (December 1898): 540–44; "Indian Hop-Pickers," *Overland Monthly* 37 (February 1891): 162–65.

50. W. P. Bonney, "Tacoma Deals with the Chinese," *History of Pierce County, Washington* (Chicago: Pioneer Historical Publishing Company, 1927), 1:451–73.

51. Carmen Olson, "Hop Growing Industry," *Issaquah Press*, November 11, 1954.

52. "Ready to Pick the Aromatic Hop," *West Side Enterprise*, September 6, 1904.

53. "Irvine's Cash Grocery," *Independence Enterprise*, August 22, 1895.

54. "James F. Clark, Mayor of Hop City, The Largest Single Hop Grower in the Country," *Grip's Valley Gazette* 1, no. 2 (October 1893): 2–5.

55. An Otsego County Hop-Grower, "The Culture of the Hop—No. 2," *Cultivator* 1, no. 3 (March 1853): 83; H.C., "Hops," *Genesee Farmer and Gardener's Journal* 5, no. 33 (August 1853): 259.

56. Ibid.

57. "James F. Clark"; James Fenimore Cooper, *Reminiscences of Mid-Victorian Cooperstown and a Sketch of William Cooper* (Cooperstown, N.Y.: Otsego County Historical Society, 1936), 36–37.

58. "Hop Dig," *Waterville Journal*, September 29, 1855; "Millions in Farming," *Waterville Times*, September 9, 1873.

59. "James F. Clark."

60. Eames, "The Hop Industry in Mendocino," 490; and "In Hop-Picking Time," 34.

61. *Semi-Weekly Appeal*, August 7, 1913.

62. George L. Bell, "The Wheatland Hop-Fields' Riot," *Outlook* 107, no. 3 (May 1914): 118–23.

63. C. H. Parker, "The California Casual and His Revolt," *Quarterly Journal of Economics* 30 (November 1915): 111.

64. Carlton H. Parker, "The Wheatland Riot and What Lay Back of It," *Survey* 31 (March 1914): 768–70.

65. Ibid.

66. For more information on the IWW, see Ralph Chaplin, *Wobbly: The Rough and Tumble Story of an American Radical* (Chicago: University of Chicago Press, 1948), and Paul F. Brissenden, *The I. W. W.: A Story of American Syndicalism* (New York: Russell and Russell, 1957).

67. "Serious Riot in Wheatland Hop Fields Sunday," *Marysville Evening Democrat*, August 4, 1913.

68. Ervin C. Moraign, "After the Riot at Wheatland" (Sacramento State College, 1959, typescript), 40.

69. "Independence Hop Festival, Aug. 29, 30, 31," *Pacific Hop Grower* 3, no. 4 (August 1935): 6. However, after the war, the nature of the industry had changed so dramatically that these entertainment functions were not resuscitated. Frances Blakely, "Oregon's Annual Hop Harvest," *Oregon Journal*, October 27, 1946, in the "Pacific Parade Magazine."

70. "Recreation in the Oregon Hop Fields," *Playground* 17, no. 12 (March 1924): 645, 646, 670.

71. Paul H. Landis, "The Hop Industry: A Social and Economic Problem," *Economic Geography* 22, no. 2 (January 1939): 85–94; Carl F. Reuss, Paul H. Landis, and Richard Wakefield, *Migratory Farm Labor and the Hop Industry on the Pacific Coast with Special Applications to Problems of the Yakima Valley, Washington*, Rural Sociology Series in Farm Labor, no. 3 (Pullman: State College of Washington, Agricultural Experiment Station, 1938).

*5. Hop Kilns, Hop Houses, and Hop Driers*

1. Solon Robinson, "Hops," *Plow: A Monthly Journal of Rural Affairs* (September 1852): 267.

2. Reginald Scot, *A Perfite Platforme of a Hoppe Garden* (1574; reprint ed., New York: DeCapo Press, 1973), 38–41.

3. Ibid.

4. Scot's design was repeated in a book by Richard Bradley, *The Riches of a Hop Garden Explained*, published in 1729. See also George Clinch, *English Hops: A History of Cultivation and Preparation for the Market from the Earliest Times* (London: McCorquodale, 1919), 85; Robin A. E. Walton, *Oasts in Kent, Sixteenth–Twentieth Century* (Maidstone, Kent, U.K.: Christine Swift Bookshop, 1985), 8.

5. R. W. Brunskill, *Traditional Farm Buildings of Britain* (London: Victor Gollancz, 1982), 95–97.

6. Edward J. Lance, *The Hop Farmer; or, A Complete Account of Hop Culture, Embracing Its History, Law, and Uses: A Theoretical and Practical Inquiry into an Improved Method of Culture, Founded on Scientific Principles* (London: Joseph Rogerson, 1838), 146–47. The attempt to employ more than one drying floor would be repeated several times in succeeding years, as will be shown.

7. "Culture of Hops in Massachusetts," *Yankee Farmer and Newsletter* 4, no. 30 (July 1838): 233.

8. Donald B. Marti, "Agricultural Journalism and the Diffusion of Knowledge: The First Half-Century in America," *Agricultural History* 54, no. 1 (January 1980): 28–37. The author notes, for example, that the early nineteenth-century issues of the *Massachusetts Agricultural Journal* were filled with material reprinted from English sources.

9. A number of agricultural experts in the United States are listed as having been consulted. The title page indicates that the work was being distributed in New York, Boston, Salem, New Haven, Newburyport, Portsmouth, Portland, Baltimore, Washington, Georgetown, Alexandria, Fredericksburg, Richmond, Petersburg, Charleston, Savannah, Augusta, Pittsburg, and Lexington. Abraham Rees, ed., *The New Cyclopedia; or, Universal Dictionary of Arts, Sciences, and Literature* (Philadelphia: Samuel F. Bradford, 1818), vol. 1, title page; Robert Collison, *Encyclopaedias: Their History Throughout the Ages* (New York: Hafner, 1964), 109–10.

10. Rees, *The New Cyclopedia*, vol. 24, n.p.

11. Ibid.

12. Walton, *Oasts in Kent*, 75.

13. Brunskill, *Traditional Farm Buildings of Britain*, 97.

14. See, for example, "The Culture of Hops," *Genesee Farmer* 4, no. 1 (January 1834): 3.

15. John Claudius Loudon, "Malt-Houses, Kilns, Hop-Oasts, etc.," *An Encyclopedia of Cottage, Farm, and Villa Architecture and Furniture* (London: Longman, Rees, Orme, Brown, Green, and Longman, 1836), 595–99.

16. Lance, *The Hop Farmer*, title page.

17. Ibid., 141–42.

18. Ibid., 146–47.

19. Ibid., 205.

20. Israel Thorndike, "Further Information on the Curing of Hops," *New England Farmer* 2, no. 5 (August 1823): 38. The letter, dated July 4, 1823, was copied from the *American Farmer*.

21. William Blanchard, Jr., "For the New England Farmer," *New England Farmer* 2, no. 7 (September 1823): 53.

22. Ibid.

23. *Transactions of the Society for the Promotion of Agriculture, Arts, and Manufactures* (Albany, N.Y.: Charles R. and George Webster, 1801), 1:149. Malt kilns were more commonplace throughout the nineteenth century.

24. "Gathering and Curing Hops," *Cultivator: A Monthly Publication, Devoted to Agriculture* 1, no. 1 (March 1834): 10, 11.

25. "Culture of Hops," *Genesee Farmer and Gardener's Journal* 3, no. 7 (February 1833): 54;

"On the Culture of Hops," *Genesee Farmer* 4, no. 46 (November 1834): 363.

26. Lincoln Cummings, "Hop Culture: Drying, & C.," *Genesee Farmer* 8, no. 7 (July 1847): 153. Cummings, a native of Augusta Township, Oneida County, New York, indicated his kiln was a banked structure, and noted that "the gable ends are boarded up, leaving a door in one end, and a window in each end to let off the steam."

27. On a homestead one and a quarter miles east of Pratts Hollow, Michael Tuke, a Methodist minister and farmer, planted hops on his farm about 1848. Shortly thereafter he built his first kiln, a small building about ten feet square with a drying floor eight feet above the charcoal fire on the floor. Arthur J. Tooke, "History of the Tooke Homestead" (Ithaca, N.Y.: Department of Manuscripts and Archives, Cornell University, 1963), 3. See also *Waterville Times*, September 3, 1868.

28. Ezra Leland, "Hops," *Transactions of the New-York State Agricultural Society, Together with an Abstract of the Proceedings of the County Agricultural Societies, and the American Institute (of the City of New York)* (Albany, N.Y.: E. Mack, 1845), 447–48.

29. *Waterville Times*, August 20, 1868.

30. Photographs taken before the destruction of the storage barn show that its first-story siding, next to the kiln, was not the same as that above. This could be taken to indicate that the first story was originally open, in the manner described above.

31. See, for example, "Hops! Hops! Hops!," an advertisement for hop stoves by J. Cross, of Morrisville, in the *Madison Observer*, August 15, 1848.

32. [Henry] Colman, "The Cultivation of Hops," *Cultivator* 4, no. 3 (March 1847): 83. The description copies "Buckland's Report of Kent." Another series described the farm of Mr. Neame, of Kent. "The Kilns for Hop Drying," *Cultivator* 7, no. 11 (November 1859): 348.

33. The myth began when the artist, anti-quarian, and lecturer Marion Nicholl Rawson, in her book *Of the Earth Earthy* (New York: E. P. Dutton, 1937), 132, wrote that the "first kiln of James Cooledge, built in 1815, is still standing in good condition for us to study. It is of solid cobblestone and stands close to the road, outliving many of its contemporaries." True, the kiln was standing on property owned by Cooledge, and he is known to have taken his first crop of hops to market in the fall of 1816. There is no indication, however, that this kiln was his first. The date is highly suspect because cobblestone construction requires a considerable quantity of hydraulic cement. Indeed, one of the principal reasons for the prevalence of cobblestone buildings in upstate New York is that Canvass White, an assistant engineer working on the Erie Canal, discovered a source of natural hydraulic cement near Sullivan, in Madison County, in 1818. White began experimentation that led to a product he patented in 1820, and, because he sold his rights to New York State, the technology became generally available soon afterward. Thus, even if Cooledge had a knowledge of the oast form in 1815, it is highly unlikely that he would have discovered a local source of hydraulic cement before White did and that his discovery would go unnoticed. See Harley J. McKee, "Canvass White and Natural Cement," *Journal of the Society of Architectural Historians* 20, no. 4 (1961): 195; Robert W. Leslie, *History of the Portland Cement Industry in the United States* (New York: Arno Press, 1972), 2. The Canvass White Papers are in the Department of Manuscripts and Archives, John M. Olin Library, Cornell University, Ithaca, New York.

34. Another example, owned by Fred Marshall, stands on the east side of Valley Mills Road, two-tenths of a mile north of Haslauer Road, in Madison County. It reportedly was built in 1867. "Hops Once Ruled Area," *Oneida Daily Dispatch*, July 3, 1977. Although the cowl was considered an improvement in kiln design and enjoyed widespread popularity throughout the late nineteenth century, by the mid-1870s it

was called into question for allowing too much draft. See "The Hop Harvest," *Oregon Cultivator* 3, no. 30 (September 1876): 3, which quotes the *Pacific Rural California Press*, which, in turn, abstracts the remarks made by J. E. Morrow of Oneida County at a meeting of the central New York hop growers.

35. Lincoln Cummings, "Hop Culture: Drying, &c.," *Genesee Farmer* 8, no. 7 (July 1847): 158.

36. An Otsego Hop-Grower, "Culture of Hops, no. 3," *Country Gentleman* 1, no. 10 (March 1853): 146.

37. William Blanchard, "Hop Culture," *Prairie Farmer* 19, no. 19 (May 1867): 316. Blanchard believed that the massive stone kiln produced a better-quality hop. "Cooking has been greatly economized by 'ranges' and 'stoves'; but, with all their modern improvements, who, that is 65 years of age, pretends that a beef steak can be so well cooked in them as our mothers and grandmothers used to do it, on an old-fashioned gridiron, over a good bed of hot maple coals. The same principle applies to drying and curing hops. I think the truth of this doctrine is too well established to require an argument."

38. Paul E. Sprague, "The Origins of Balloon Framing," *Journal of the Society of Architectural Historians* 40, no. 4 (December 1981): 311–19. Sprague has also dealt at length with this development in "Chicago Balloon Frame," in *The Technology of Historic American Buildings*, ed. H. Ward Jandl (Washington, D.C.: Foundation for Preservation Technology, 1983), 35–61. For additional information about the two most prominent upstate New York proponents of the balloon frame, see Michael A. Tomlan, "Popular and Professional American Architectural Literature in the Late Nineteenth Century" (Ph.D. diss., Cornell University, 1983), 131–43. For a thorough treatment of heavy timber construction, see John I. Rempel, *Building with Wood* (Toronto: University of Toronto Press, 1980), 91–260.

39. J. W. Clarke, "Hops: Picking and Curing, II," *Cultivator* 3, no. 6 (June 1860): 176.

40. This section draws heavily on the author's earlier work, "Hop Houses in Central New York, with Guidelines for Their Identification and Evaluation," an unpublished report prepared for the Madison County Historical Society, Oneida, New York, under contract to the New York Office of Parks, Recreation and Historic Preservation, Albany, New York, in 1983.

41. Andrew Samuel Fuller, "Hop Culture," in *Hop Culture: Practical Details, from the Selection and Preparation of the Soil, and Setting and Cultivation of the Plants, to Picking, Drying, Pressing, and Marketing the Crop*, ed. Andrew Samuel Fuller (New York: Orange Judd, 1865), 1–2.

42. Heman C. Collins, "No. I," in *Hop Culture*, ed. Fuller, 7–12.

43. E.O.L. "No. II," in *Hop Culture*, ed. Fuller, 14–16.

44. A. F. Powley, "No. IV," in *Hop Culture*, ed. Fuller, 21–22.

45. Albert W. Morse, "No. V," in *Hop Culture*, ed. Fuller, 25.

46. H. H. Potter, "Hop Houses and Hop Boxes," *Prairie Farmer*, n.s. 19, no. 12 (March 1867): 183.

47. D. B. Rudd and E. O. Rudd, *The Cultivation of Hops, and Their Preparation for Market as Practiced in Sauk County, Wisconsin* (Reedsburg, Wis.: Privately published, 1868).

48. "Hop Craze in Sauk County," in *The History of Sauk County, Wisconsin* (Chicago: Western Historical Company, 1880), 619.

49. Harry Ellsworth Cole, ed., *A Standard History of Sauk County, Wisconsin* (Chicago and New York: Lewis, 1918), 1:57.

50. "Picking and Drying Hops," *Wisconsin Mirror*, August 5, 1868; "The Farmer's Department," *Daily Appeal*, September 19, 1861.

51. The door to the upper story has been covered with vertical boards, and a new roof was installed in 1974.

52. "Hon. Ephraim Beaumont," in *Portrait*

and *Biographical Record of Waukesha County, Wisconsin* (Chicago: Excelsior, 1894), 728.

53. Charles F. Calkins and William G. Laatsch, "The Hop Houses of Waukesha County, Wisconsin," *Pioneer America* 9, no. 2 (December 1977): 180–207. The authors assumed that because (1) the size of the yards in Waukesha County was generally small, and (2) the Beaumont kiln, another kiln that they were able to identify in Waukesha County, and a few foundations measured no larger than about twenty feet square, it followed that most of the kilns built in the area lacked their attendant cooling and baling barns. The authors go on to state that Wisconsin hop growers seemed to ignore New York State models and instead to follow other schemes found in Great Britain. Although the observations are generally correct, as has been shown, it is dangerous to draw such conclusions on so little evidence.

54. "Hop-Houses Burned," *Independent,* September 24, 1867. The column notes that two hop houses recently lost had been insured for $2,500 and $2,000, respectively, and that this made six that had been lost in the county during the present season. See also "Hops and the Panic of 1868," in *The History of Sauk County, Wisconsin,* 366; John Truc, "Hop Days in Sauk County," in *A Standard History of Sauk County, Wisconsin* (Chicago and New York: Lewis, 1918), 95.

55. "Edward France's Improved Hop-Drier," in *Hop Culture,* ed. Fuller, advertisement, inside back cover.

56. *Annual Report of the Commissioner of Patents for the Year 1868* (Washington, D.C.: U.S. Government Printing Office, 1868), 1:702.

57. "Best Hop Yard and Hop House," *Wisconsin Mirror,* September 1, 1869; "Self-Dumping Hop Floor," *Wisconsin Mirror,* September 22, 1869. One such example was built on the farm of L. S. Dickens, about five miles west of Kilbourn City.

58. *Annual Report of the Commissioner of*

*Patents for the Year 1869* (Washington, D.C.: U.S. Government Printing Office, 1869), 3:338.

59. Daniel Flint, "Hop Growing in California," *Pacific Rural Press* 21, no. 24 (June 1881): 416.

60. *Annual Report of the Commissioner of Patents for the Year 1868* (Washington, D.C.: U.S. Government Printing Office, 1868), 4:982.

61. "The Farmer's Department," *Daily Appeal,* September 19, 1861. Previously all hops "raised here have been dried by the sun." In this context it should be noted that no evidence has come to light indicating that hop growers on the Pacific Coast were looking over their shoulders at the rising fruit-drying industry. In fact, if the contemporary literature is any indication, because hop kilns were relatively sophisticated, any transfer of technology would likely have been to the benefit of the fruit growers. See, for example, "The Plummer Fruit Dryer," *Napa Register,* July 20, 1878. This patented machine was sold in two models, one for factory and another for family use.

62. William M. Haynie, "Hop Culture in California," in *Transactions of the California State Agricultural Society During the Years 1866 and 1867* (Sacramento: D. W. Gelwicks, 1868), 448. The article not only describes the kiln then commonly in use, but also Haynie's own invention, an improved drying process in which the hops were loaded in long boxes fitted with wire-cloth tops and bottoms, then run into the kiln on a railroad track. It was claimed that removing the boxes and turning them allowed the hops to dry more quickly and evenly.

63. Sucheng Chan, *This Bitter-Sweet Soil: The Chinese in California Agriculture, 1860–1910* (Berkeley and Los Angeles: University of California Press, 1986), 248. Chan notes that in some cases the leases stipulated that the Chinese had to build the hop houses and pay one-half the fire insurance.

64. "Hints on Hop Growing, no. 7," *Pacific Rural Press* 8, no. 3 (July 1874): 40.

65. "Hop Houses," *Pacific Rural Press* 18,

no. 15 (October 1879): 73. The biggest difficulty seems to have been that there simply were not enough kilns. "With the exception of the Messrs. Bird of Santa Clara County and possibly a few others, there is hardly a yard in the state with sufficient [drying] capacity." Editorial, *Pacific Rural Press* 9, no. 20 (May 1875): 2.

66. Herbert Myrick, *The Hop: Its Culture and Cure, Marketing and Manufacture* (New York: Orange Judd, 1899), 185.

67. Ibid., 247.

68. Ibid.

69. "The Hop Industry in Mendocino," *Pacific Rural Press* 42, no. 24 (December 1891): 490.

70. "First Hop Yard," *Eugene Daily Guard*, September 9, 1899. What were believed to be the two earliest hop houses in Oregon were still standing in 1899, when the *Eugene Daily Guard* urged that photographs be taken of them. If this record was created, however, it has never come to light. There was no mention of the materials employed in the buildings. See also William Wells, "Hop Culture in Oregon," *Willamette Farmer*, May 7, 1870.

71. "First Washington Crop," *Washington Standard*, September 23, 1887. Ezra Meeker, *Hop Culture in the United States, Being a Practical Treatise on Hop Growing in Washington Territory* (Puyallup, Washington Territory: E. Meeker, 1883), 8.

72. Meeker, *Hop Culture in the United States*, 49–50.

73. "Hon. L. F. Thompson: Sketch of His Life and Character" (Washington State Historical Society, n.d., manuscript). Although Thompson was born in Jamestown, New York, no indication has come to light that he had any particular prior knowledge of or interest in hops. His first occupation was that of a hotel keeper in Steilacoom, Washington, until 1860, when he relocated in the Puyallup Valley.

74. "Hop Fields of Puyallup and White River," *West End* 10, no. 11 (November 1884): 346.

75. "Hops," *Weekly Pacific Tribune*, October 9, 1874.

76. "A Puyallup Hop Farm," *Pacific Daily Tribune*, August 30, 1877.

77. "Hop Picking," *Washington Standard*, September 24, 1886.

78. Ross's home was near the present site of the Experiment Station. W. P. Bonney, *History of Pierce County, Washington* (Chicago: Pioneer Historical Publishing Company, 1927), 1:392.

79. "Hop Fields of Puyallup and White River," *West Shore* 10, no. 11 (November 1884): 347.

80. Ibid.

81. News item, *Washington Standard*, September 15, 1882.

82. S. R., J. C., and J. H. Templeton had patented a similar device in 1877.

83. On at least one occasion, Meeker is known to have provided plans for hop kilns to a correspondent. See Norman Rideout to Ezra Meeker, May 25, 1893, in "Hop Correspondence (October 1890–June 1893)," Box 3, Folder 1, Ezra Meeker Papers, Washington State Historical Society, Tacoma.

84. "In the Hop Fields," *Northwest Pacific Farmer*, August 30, 1894, 1.

85. "Another Method of Drying Hops," *Northwest Pacific Farmer*, July 26, 1894, 1.

86. It should be noted that next to the hop field, in the foreground of the photograph, is a vegetable garden for the convenience of the hop pickers, who were camped in tents immediately to the right.

87. Hopkins' patent fan process, using double floors and a suction fan at the top of the kiln, was mentioned by Meeker in *Hop Culture in the United States*, 27.

88. "The Drying of Hops," *Pacific Hop Grower* 3, no. 7 (November 1935): 3.

89. C. J. Hurd, "Fans in Hop Dryers," *Pacific Hop Grower* 2, no. 9 (January 1934): 3, 6.

90. "E. C. Horst Uses Fans," *Pacific Hop Grower* 3, no. 2 (June 1935): 5.

91. "Water Boosts Hop Yields," *Oregon Farmer*, July 13, 1933, 2.

92. "An Automatic Baler," *Pacific Hop Grower* 2, no. 3 (July 1934): 7.

93. "Farm Betterments," *Oregon Farmer*, September 7, 1933, 10.

94. "New Hop Houses Rise," *Oregon Hop Grower* 1, no. 3 (June 1933): 7.

95. "New Airblast Kilns Erected at Moxee City," *Oregon Hop Grower* 1, no. 4 (July 1933): 4. A number of new drying facilities were built in the Yakima Valley at this time. See "Farm Betterments," *Oregon Farmer,* December 14, 1933, 10, and April 5, 1934, 14.

96. "Railroad Company to Build Fireproof Warehouse," *Pacific Hop Grower* 3, no. 3 (July 1935): 5.

97. Karl H. Lehman, "Hops," in *Madison County Today* (Oneida Castle, N.Y.: Karl H. Lehman, 1943), 54.

98. C. Ivan Branton, *A Hop Drier for Oregon Farms* (Corvallis: Oregon State College, Agricultural Experiment Station, 1950); Sue Ehrstine, "Sunshine and Beer," *Quest* 16, no. 1 (Winter 1982): 17–18.

*Appendix*

1. Sereno Edwards Todd, "Hints About Hop-Houses," in *Annual Report of the American Institute, of the City of New York, for the Year 1867–68* (Albany, N.Y.: Chas. Van Benthuysen and Sons, 1868), 184–86.

# BIBLIOGRAPHY

*Books and Bulletins*

Adams, R. L. *Seasonal Labor Requirements of California Crops.* University of California College of Agriculture Bulletin no. 623. Berkeley: University of California, Berkeley, 1938.

*And When They're Gone.* Eugene, Ore.: Lane County Department of Employment and Training, 1980.

Anderson, W. A. *Social Change in a Central New York Community.* Cornell University Agricultural Experiment Station Bulletin no. 907. Ithaca, N.Y.: Cornell University, 1957.

Anglo-American Council on Productivity. *The Hop Industry.* New York: Productivity Team, 1951.

*Annual Report of the American Institute, of the City of New York for the Years 1865–66.* Albany, N.Y.: C. Wendell, 1866.

*Annual Report of the Commissioner of Patents for the Year 1853: Agriculture.* Washington, D.C.: A. O. P. Nicholson, Printer, 1854.

*Annual Report of the Commissioner of Patents for the Year 1868.* Washington, D.C.: U.S. Government Printing Office, 1868.

*Annual Report of the Commissioner of Patents for the Year 1869.* Washington, D.C.: U.S. Government Printing Office, 1869.

Arnold, John P., and Frank Penman. *History of the Brewing Industry and Brewing Science in America, Prepared as Part of a Memorial to the Pioneers of American Brewing Science, Dr. John E. Siebel and Anton Schwarz.* Chicago: G. L. Peterson, 1933.

Bancroft, Hubert Howe. *The History of California.* San Francisco: History Company, 1884–90.

Baron, Stanley. *Brewed in America: A History of Beer and Ale in the United States.* Boston: Little, Brown, 1962.

Beers, Frederick W. *Atlas of Otsego County, New York.* New York: Frederick W. Beers, 1868.

Bidwell, Percy Wells, and John I. Falconer. *History of Agriculture in the Northern United States.* Washington, D.C.: Carnegie Institution, 1925.

Blocker, Jack S., Jr., ed. *Alcohol, Reform, and Society.* Westport, Conn.: Greenwood Press, 1979.

Blodgett, F. M. *Hop Mildew.* Cornell University Agricultural Experiment Station Bulletin no. 328. Ithaca, N.Y.: Cornell University, 1913.

Bonney, W. P. *History of Pierce County, Wash-*

*ington*. Vol. 1. Chicago: Pioneer Historical Publishing Company, 1927.

Branton, C. Ivan. *A Hop Drier for Oregon Farms*. Corvallis: Oregon State College, Agricultural Experiment Station, 1950.

Brissenden, Paul F. *The I. W. W.: A Story of American Syndicalism*. New York: Russell and Russell, 1957.

Brooks, S. N., C. E. Horner, and S. T. Likens. *Hop Production*. Agricultural Information Bulletin no. 240. Washington, D.C.: U.S. Government Printing Office, 1961.

Brunskill, R. W. *Traditional Farm Buildings of Britain*. London: Victor Gollancz, 1982.

Bull, Donald, and Manfred Friedrich. *The Register of the United States Breweries, 1876–1976*. Trumbull, Conn.: Privately published, 1976.

Burgess, Abraham Hale. *Hops: Botany, Cultivation, and Utilization*. London: Leonard Hill; New York: Interscience Publishers, 1964.

Carrier, Lyman. *The Beginnings of Agriculture in America*. New York: McGraw-Hill, 1923.

Chan, Sucheng. *This Bitter-Sweet Soil: The Chinese in California Agriculture, 1860–1910*. Berkeley and Los Angeles: University of California Press, 1986.

Chaplin, Ralph. *Wobbly: The Rough and Tumble Story of an American Radical*. Chicago: University of Chicago Press, 1948.

Chapman, Alfred Chaston. *The Hop and Its Constituents: A Monograph on the Hop Plant*. London: Brewing Trade Review, 1905.

Clinch, George. *English Hops: A History of Cultivation and Preparation for the Market from the Earliest Times*. London: McCorquedale, 1919.

Coe, Robert, Jr., Miriam Griffith, Helen Hudson Stineman, and Juanita McCurry Neyens. *Wheatland, 1874–1974*. Wheatland, Calif.: The Wheatland Historical Society, 1974.

Colburn, William P. *The History of Milford*. Concord, N.H.: Rumford Press, 1901.

Cole, Harry Ellsworth, ed. *A Standard History of Sauk County, Wisconsin*. Chicago and New York: Lewis, 1918.

*Collections of the History of Albany, from Its Discovery to the Present Time*. Vol. 1. Albany, N.Y.: J. Munsell, 1865.

Cooper, James Fenimore. *Reminiscences of Mid-Victorian Cooperstown and a Sketch of William Cooper*. Cooperstown, N.Y.: Otsego County Historical Society, 1936.

Day, Clarence Albert. *Farming in Maine, 1860–1940*. Orono: University of Maine Press, 1963.

DeBow, J. D. B., ed. *Statistical View of the United States . . . Being a Compendium of the Seventh Census*. Washington, D.C.: A. O. P. Nicholson, 1854.

———, supt. *The Seventh Census of the United States: 1850*. Washington, D.C.: Robert Armstrong, 1853.

De Loach, D. B. *Outlook for Hops from the Pacific Coast*. Washington, D.C.: U.S. Bureau of Agricultural Economics, 1948.

Dickinson, Rudolphus. *A Geographical and Statistical View of Massachusetts Proper*. Greenfield, Mass.: Denio and Phelps, 1813.

Drake, Samuel Adams. *History of Middlesex County, Massachusetts, Containing Carefully Prepared Histories of Every City and Town in the County*. Boston: Estes and Lauriat, 1880.

Dunlap, Thomas R. *DDT: Scientists, Citizens, and Public Policy*. Princeton, N.J.: Princeton University Press, 1981.

*Eighty Years' Progress of the United States: A Family Record of American Industry, Energy, and Enterprise*. Hartford, Conn.: L. Stebbins, 1868.

Emmons, Ebenezer. *Agriculture of New York*. Albany, N.Y.: C. Benthuysen, 1846–54.

*Fifteenth Annual Report of the New York State Agricultural Society, for the Year 1890*. Albany, N.Y.: James B. Lyon, 1891.

Filmer, Richard. *Hops and Hop Picking*. Aylesbury, Bucks, England: Shire Publications, 1986.

Flint, Daniel. United States Department of Agriculture, *Hop Culture in California*. Farmers' Bulletin no. 115. Washington, D.C.: N. p., 1900.

Fogg, Alonzo J., comp. *The Statistics and Gazetteer of New Hampshire*. Concord, N.H.: D. L. Guernsey, 1874.

French, J. H. *Historical and Statistical Gazeteer of New York State*. Syracuse, N.Y.: R. P. Smith, 1860.

Fuller, Andrew S., ed. *Hop Culture: Practical Details from the Selection and Preparation of the Soil, and Setting and Cultivation of the Plants, to Picking, Drying, Pressing, and Marketing the Crop*. New York: Orange Judd, 1865.

Glassie, Henry. *Pattern in the Material Folk Culture of the Eastern United States*. Philadelphia: University of Pennsylvania Press, 1968.

Gross, Emanuel. *Hops, in Their Botanical, Agricultural, and Technical Aspect and as an Article of Commerce*. London: Scott, Greenwood, 1900.

Hedrick, Ulysses P. *History of Agriculture in the State of New York*. New York: New York State Historical Association, 1966.

*History of Bedford, New-Hampshire, Being Statistics, Compiled on the Occasion of the One Hundredth Anniversary of the Incorporation of the Town*. Boston: Alfred Mudge, 1851.

*History of Oneida County, New York, from 1700 to the Present Time*. Vol. 2. Chicago: S. J. Clarke, 1912.

*History of Otsego County, New York, with Illustrations and Biographical Sketches of Some of Its Prominent Men and Pioneers*. Philadelphia: Everts and Fariss, 1878.

Hollands, Harold F., Edgar B. Hurd, and Ben H. Purbois. *Economic Conditions and Problems of Agriculture in the Yakima Valley, Washington*. Part 4, *Hop Farming*. Pullman, Wash.: Washington Agricultural Experiment Station, 1942.

Holmes, George K., comp. *Hop Culture of the United States, 1790–1911*. U.S. Department of Agriculture, Bureau of Statistics, Circular no. 35. Washington, D.C.: U.S. Government Printing Office, 1912.

Hurd, D. Hamilton. *History of Middlesex County, Massachusetts, with Biographical Sketches of Many of Its Pioneers and Prominent Men*. Vol. 1. Philadelphia: J. W. Lewis, 1890.

Ibanez, Reuben, ed. *Historical Bok Kai Temple in Old Marysville, California*. Marysville, Calif.: Marysville Chinese Community, 1967.

Jackson, James R. *History of Littleton, New Hampshire, in Three Volumes*. Cambridge, Mass.: University Press, 1905.

Jones, Pomroy. *Annals and Recollections of Oneida County*. Rome: N.Y.: Privately published, 1851.

*Journal of the Assembly of the State of New York at Their Forty-third Session*. Albany, N.Y.: J. Buel, 1820.

Kuhlman, G. W., and R. E. Fore. *Cost and Efficiency in Producing Hops in Oregon*. Agricultural Experiment Station Bulletin no. 364. Corvallis: Oregon State College, 1939.

Lance, Edward J. *The Hop Farmer; or, A Complete Account of Hop Culture, Embracing Its History, Laws, and Uses: A Theoretical and Practical Inquiry into an Improved Method of Culture, Founded on Scientific Principles*. London: Joseph Rogerson, 1838.

Leslie, Robert W. *History of the Portland Cement Industry in the United States*. New York: Arno Press, 1972.

Loudon, John Claudius. *An Encyclopedia of Agriculture, Comprising the Theory and Practice of the Valuation, Transfer, Laying Out, Improvement, and Management of Landed Property*. London: Longman, Rees, Orme, Brown, Green and Longman, 1831.

———. *An Encyclopedia of Cottage, Farm, and Villa Architecture and Furniture*. London: Longman, Rees, Orme, Brown, Green and Longman, 1836.

Meeker, Ezra. *Hop Culture in the United*

States, Being a Practical Treatise on Hop Growing in Washington Territory, from the Cutting to the Bale . . . to Which is Added an Exhaustive Article from the Pen of A. W. Lawrence, Esq., Waterville, N.Y., on Hop Raising in New York State. Puyallup, Washington Territory: E. Meeker, 1883.

Merritt, Eugene. Hops in Principal Countries: Their Supply, Foreign Trade, and Consumption, with Statistics of Beer Brewing. U.S. Department of Agriculture, Bureau of Statistics. Washington, D.C.: N. p., 1907.

Myrick, Herbert. The Hop: Its Culture and Cure, Marketing and Manufacture. New York: Orange Judd, 1899.

New York Secretary of State. Census of the State of New York for 1865. Albany, N.Y.: C. Van Benthuysen and Sons, 1865.

Nickles, Napoleon. Le Houblon. Paris: Librarie Agricole de la Maison Rustique, 1847.

O'Callaghan, Edmund Bailey. History of New Netherland; or, New York Under the Dutch. New York: D. Appleton, 1846.

―――, ed. Documents Relating to the Colonial History of the State of New York. Albany, N.Y.: Weed, Parsons, 1856.

―――. Laws and Ordinances of New Netherland, 1638–1674. Albany, N.Y.: Weed, Parsons, 1868.

Ockey, W. C., Dallas W. Smythe, and F. R. Wilcox. Statistics Presented with a Proposed Marketing Agreement for Hops Produced in California, Washington, and Oregon. Berkeley, Calif.: University of California Extension Service, 1935.

Portrait and Biographical Record of Waukesha County, Wisconsin. Chicago: Excelsior, 1894.

Prescott, Henry Paul. Strong Drink and Tobacco Smoke: The Structure, Growth, and Uses of Malt, Hops, Yeast, and Tobacco. New York: W. Wood, 1970.

Rawson, Marion Nicholl. Of the Earth Earthy. New York: E. P. Dutton, 1937.

Rees, Abraham, ed. The New Cyclopedia; or, Universal Dictionary of Arts, Sciences, and Literature. Vol. 1. Philadelphia: Samuel F. Bradford, 1818.

Rempel, John I. Building with Wood. Toronto: University of Toronto Press, 1980.

Reuss, Carl F., Paul H. Landis, and Richard Wakefield. Migratory Farm Labor and the Hop Industry on the Pacific Coast with Special Applications to Problems of the Yakima Valley, Washington. Rural Sociology Series in Farm Labor, no. 3. Pullman: State College of Washington, Agricultural Experiment Station, 1938.

Rorabaugh, W. J. The Alcoholic Republic, An American Tradition. New York: Oxford University Press, 1979.

Rudd, D. B., and E. O. Rudd. The Cultivation of Hops, and Their Preparation for Market as Practiced in Sauk County, Wisconsin. Reedsburg, Wis.: Privately published, 1868.

Ruggles, Samuel B. Tabular Statements from 1840 to 1870, of the Agricultural Products of the States and Territories of the United States of America. New York: Press of the Chamber of Commerce, 1874.

Russell, Howard S. A Long, Deep Furrow: Three Centuries of Farming in New England. Hanover, N.H.: University Press of New England, 1976.

Salem, Frederick William. Beer: Its History and Its Economic Value as a National Beverage. Hartford, Conn.: F. W. Salem, 1880.

Salerud, George. An Economic Study of the Hop Industry in Oregon. Corvallis: Oregon Agricultural Experiment Station, 1931.

Schull, Diantha Dow. Landmarks of Otsego County. Syracuse, N.Y.: Syracuse University Press, 1980.

Scot, Reginald. A Perfite Platforme of a Hoppe Garden. 1574. Reprint. New York: Da Capo Press, 1973.

Secomb, David F. History of the Town of Amherst, Hillsborough County, New Hampshire. Concord, N.H.: Evans, Sleeper and Woodbury, 1883.

Simmonds, Peter Lund. Hops: Their Cultiva-

tion, Commerce, and Uses in Various
Countries. A Manual of Reference for the
Grower, Dealer, and Brewer. London, N.Y.:
E. and F. N. Spon, 1877.

Smith, James H. History of Chenango and
Madison Counties, New York. Syracuse,
N.Y.: D. Mason, 1880.

Smith, John E., ed. A Descriptive and Biographical Record of Madison County, New
York. Boston: Boston History Company, 1899.

Steiner, S. S., Incorporated. Steiner's Guide to
American Hops. New York: Hopsteiner,
1973.

Stockberger, W. W. Growing and Curing of
Hops. United States Department of Agricul-
ture, Farmers' Bulletin no. 304. Washington,
D.C.: U.S. Government Printing Office, 1928.

Stockberger, W. W. The Necessity for New
Standards of Hop Valuation. United States
Department of Agriculture, Bureau of Plant
Industry, Circular no. 33. Washington, D.C.:
U.S. Government Printing Office, 1913.

The History of Sauk County, Wisconsin. Chicago: Western Historical Company, 1880.

The Portrait and Biographical Record of
Waukesha County, Wisconsin. Chicago: Excelsior, 1894.

Transactions of the California State Agricultural Society During the Year 1859. Sacramento: C. T. Botts, 1860.

Transactions of the California State Agricultural Society During the Year 1860. Sacramento: Benjamin P. Avery, 1861.

Transactions of the California State Agricultural Society During the Year 1863. Sacramento: O. M. Clayes, 1864.

Transactions of the California State Agricultural Society During the Years 1864 and 1865.
Sacramento: O. M. Clayes, 1866.

Transactions of the California State Agricultural Society During the Years 1866 and 1867.
Sacramento: D. W. Gelwicks, 1868.

Transactions of the California State Agricultural Society During the Years 1868 and 1869.
Sacramento: D. W. Gelwicks, 1870.

Transactions of the California State Agricultural Society During the Years 1870 and 1871.
Sacramento: T. A. Springer, 1872.

Transactions of the California State Agricultural Society During the Year 1875. Sacramento: State Printing Office, 1876.

Transactions of the California State Agricultural Society During the Year 1879. Sacramento: J. D. Young, 1880.

Transactions of the California State Agricultural Society During the Year 1891. Sacramento: A. J. Johnston, 1892.

Transactions of the California State Agricultural Society During the Year 1904. Sacramento: W. W. Shannon, 1905.

Transactions of the New-York State Agricul-
tural Society, Together with an Abstract of the
Proceedings of the County Agricultural So-
cieties, and the American Institute (of the
City of New York). Vol. 4, 1844. Albany, N.Y.:
E. Mack, 1845.

Transactions of the Society for the Promotion of
Agriculture, Arts and Manufactures, Vol. 1.
Albany, N.Y.: Charles R. and George Web-
ster, 1801.

Transactions of the Wisconsin State Agricultural Society. Vol. 7, 1861–68. Madison: Atwood and Rublee, 1868.

Tyrrell, Ian R. Sobering Up: From Temperence
to Prohibition in Antebellum America, 1800–
1860. Westport, Conn.: Greenwood Press,
1979.

U.S. Bureau of the Census. U.S. Census of Agriculture: 1959. Washington, D.C.: U.S. Government Printing Office, 1960.

U.S. Crop Reporting Board. Hops by States,
1915–69: Acreage, Field, Production, Disposition, Value, Stocks. Washington, D.C.:
Thomas Allen, 1841.

U.S. Department of Agriculture. Hops by
States, 1915–56. Washington, D.C.: U.S. Department of Agriculture, 1958.

U.S. Department of Agriculture, AAA Division
of Marketing and Marketing Agreements.
Preliminary Economic Statement Relating to

*Hops Produced in Oregon, Washington, and California*. Washington, D.C.: U.S. Government Printing Office, 1938.

U.S. Department of State. *Compendium of Enumeration of the Inhabitants and Statistics of the United States, as Obtained at the Department of State, from the Returns of the Sixth Census*. Washington, D.C.: U.S. Government Printing Office, 1873–1900.

U.S. Patent Office. *Subject Index of Patents for Inventions 1790–1873*. New York: Arno Press, 1976.

Vries, David Pietersz De. *Voyages from Holland to America, A.D. 1632 to 1644*. Translated by Henry C. Murphy. New York: James Lenox, 1853.

Walker, Francis A., and Charles W. Seaton, supts. *Report on the Productions of Agriculture as Returned at the Tenth Census (June 1, 1880)*. Washington, D.C.: U.S. Government Printing Office, 1883.

Walker, Francis A., supt. *A Compendium of the Ninth Census Compiled Pursuant to a Concurrent Resolution of Congress and Under the Direction of the Secretary of the Interior*. Washington, D.C.: U.S. Government Printing Office, 1872.

———. *Compendium of the Tenth Census (June 1, 1880), Compiled Pursuant to an Act of Congress Approved August 7, 1882*. Washington, D.C.: U.S. Government Printing Office, 1885.

Walton, Robin A. E. *Oasts in Kent, Sixteenth–Twentieth Century*. Maidstone, Kent, U.K.: Christine Swift Bookshop, 1985.

Whitehead, Charles. *Hop Cultivation*. London: John Murray, 1893.

Whitney, Luna M. Hammond. *History of Madison County, State of New York*. Syracuse, N.Y.: Truair, Smith, 1872.

Whorton, James. *Before Silent Spring: Pesticides and Public Health in Pre-DDT America*. Princeton, N.J.: Princeton University Press, 1974.

Wilcoxen, Charlotte. *Seventeenth Century Albany: A Dutch Profile*. Albany, N.Y.: Albany Institute of History and Art, 1981.

Wilson, Harold Fisher. *The Hill Country of Northern New England: Its Social and Economic History, 1790–1930*. New York: Columbia University Press, 1936.

Wilson, Thomas. *Picture of Philadelphia, for 1824, Containing the "Picture of Philadelphia, for 1811, by James Mease, M.D.," with All Its Improvements Since That Period*. Philadelphia: Thomas Town, 1823.

Wood, L. D. *Hops: Their Production and Uses*. Chicago: National Hops Company, 1938.


*Agricultural Journals and Newspapers*

*American Agriculturalist* (New York)
*American Farmer* (Baltimore)
*Anaheim Daily Gazette* (Anaheim, Calif.)
*California Farmer* (San Francisco)
*Country Gentleman* (New York)
*Cultivator* (Albany, N.Y.)
*Daily Appeal* (Marysville, Wis.)
*Daily Pacific Tribune* (Tacoma, Wash.)
*Eugene Weekly Guard* (Eugene, Ore.)
*Farmer's Monthly Visitor* (Concord, N.H.)
*Freeman's Journal* (Cooperstown, N.Y.)
*Genesee Farmer and Gardener's Journal* (Rochester, N.Y.)
*Independence Enterprise* (Independence, Ore.)
*Independent* (Baraboo, Wis.)
*Issaquah Press* (Issaquah, Wash.)
*Madison County Times* (Chittenango, N.Y.)
*Madison Observer* (Morrisville, N.Y.)
*Marysville Evening Democrat* (Marysville, Calif.)
*Moore's Rural New-Yorker* (Rochester, N.Y.)
*Napa Register* (Napa, Calif.)
*New England Farmer and Gardener's Journal* (Boston)
*New England Farmer and Horticultural Register* (Boston)
*New York Farmer and Horticultural Repository* (New York)

New York Tribune (New York)

Northwest Pacific Farmer (Portland, Ore.)

Oneida Daily Dispatch (Morrisville, N.Y.)

Oregon Agriculturalist and Rural Northwest (Portland)

Oregon Cultivator (Albany)

Oregon Farmer (Portland)

Oregonian (Portland)

Oregon Journal (Portland)

Pacific Hop Grower (Mt. Angel, Ore.)

Pacific Rural Press (San Francisco)

Plow: A Monthly Journal of Rural Affairs (New York)

Prairie Farmer (Chicago)

Puyallup Valley Tribune (Puyallup, Wash.)

Redwood Journal-Press Dispatch (Ukiah, Calif.)

Semi-Weekly Appeal (Sacramento)

Vermont Family Visitor (Montpelier)

Washington Standard (Olympia)

Waterville Times and Hop Reporter (Waterville, N.Y.)

Willamette Farmer (Salem, Ore.)

Wisconsin Farmer, and Northwestern Cultivator (Madison)

Wisconsin Mirror (Kilbourn City)

Working Farmer (New York)

Yakima Herald (Yakima, Wash.)

Yankee Farmer and Newsletter (Portland, Maine)


## Articles

Bell, George L. "The Wheatland Hop-Fields' Riot." Outlook 107, no. 3 (May 1914): 118–23.

Bull, W. H. "Indian Hop-Pickers on Puget Sound." Harper's Weekly 36, no. 1850 (July 1892): 545–46.

Calkins, Charles F., and William G. Laatsch. "The Hop Houses of Waukesha County, Wisconsin." Pioneer America 9, no. 2 (December 1977): 180–207.

Chaffin, A. L. "Two Weeks in Hop Harvest."

University of California Journal of Agriculture 2, no. 4 (December 1914): 150–51.

Currier, Susan Lord. "Some Aspects of Washington Hop-Fields." Overland Monthly 32 (December 1898): 540–44.

Darlington, James. "Hops and Hop Houses in Upstate New York." Material Culture 16, no. 1 (Spring 1984): 25–42.

Day, Clarence Albert. "A History of Maine Agriculture, 1604–1860." University of Maine Bulletin 56, no. 11 (April 1954): 126, 130.

Eames, Ninetta. "In Hop-Picking Time." Cosmopolitan 16, no. 1 (November 1893): 27–36.

Freeman, Otis W. "Hop Industry of the Pacific Coast States." Economic Geography 19, no. 3 (April 1936): 155–63.

Hadley, Edgar M. "Hops in the United States." Geographical Bulletin 2, no. 1 (April 1971): 37–51.

Hill, Joseph A. "The Oregon Hop Industry." Scientific American Supplement 51, no. 1317 (March 1901): 21113.

"Indian Hop-Pickers." Overland Monthly 17 (February 1891): 162–65.

"James F. Clark, Mayor of Hop City, The Largest Single Hop Grower in the Country." Grip's Valley Gazette 1, no. 2 (October 1893): 2–5.

Johnston, James F. W. "The Hop and Its Substitutes, from 'The Chemistry of Common Life.'" Working Farmer 7, no. 4 (June 1855): 91.

Kupperman, Karen Ordahl. "Climate and Mastery of the Wilderness in Seventeenth-Century New England." In Seventeenth Century New England, edited by David D. Hall and David Greyson Allen, 3–37. Boston: Colonial Society of Massachusetts, 1984.

Landis, Paul H. "The Hop Industry: A Social and Economic Problem." Economic Geography 22, no. 2 (January 1939): 85–94.

"Machine-Picked vs. Hand-Picked Hops," Scientific American Supplement 68, no. 1770 (December 1909): 360–61.

MacLean, Annie Marion. "With Oregon Hop

Pickers." *American Journal of Sociology* 15, no. 1 (July 1909): 83–95.

Marti, Donald B. "Agricultural Journalism and the Diffusion of Knowledge: The First Half-Century in America." *Agricultural History* 54, no. 1 (January 1980): 28–37.

McKee, Harley J. "Canvass White and Natural Cement." *Journal of the Society of Architectural Historians* 20, no. 4 (1961): 195.

Miller, Elbert, and Richard M. Highsmith. "The Hop Industry of the Pacific Coast." *Journal of Geography* 49, no. 2 (1950): 63–77.

Nelson, Herbert B. "The Vanishing Hop-Driers of the Willamette Valley." *Oregon Historical Quarterly* 64, no. 3 (September 1963): 257–71.

Newman, Doug. "History of Hops Growing in Lane County." *Lane County Historian* 28, no. 3 (Fall 1983): 70–76.

Parker, Carlton H. "The California Casual and His Revolt." *Quarterly Journal of Economics* 30 (November 1915): 111–26.

———. "The Wheatland Riot and What Lay Back of It." *Survey* 31 (March 1914): 768–70.

Parsons, James J. "Hops in Early California Agriculture." *Agricultural History* 14, no. 3 (1940): 110–16.

"Recreation in the Oregon Hop Fields." *Playground* 17, no. 12 (March 1924): 645–46, 670.

Rumney, Thomas. "Agricultural Production Locational Stability: Hops in New York State During the Nineteenth Century," *Kansas Geographer* 18 (1983): 5–16.

———. "The Hops Boom in Nineteenth Century Vermont." *Vermont History* 56, no. 1 (Winter 1988): 36–41.

Sprague, Paul E. "Chicago Balloon Frame." In *The Technology of Historic American Buildings*, edited by H. Ward Jandl, 35–61. Washington, D.C.: Foundation for Preservation Technology, 1983.

———. "The Origins of Balloon Framing." *Journal of the Society of Architectural Historians* 40, no. 4 (December 1981): 311–19.

Thirsk, Joan. "Patterns of Agriculture in Seventeenth-Century England." In *Seventeenth Century New England*, edited by David D. Hall and David Greyson Allen, 37–54. Boston: Colonial Society of Massachusetts, 1984.

Turner, Andrew Jackson. "The History of Fort Winnebago." In *Collections of the State Historical Society of Wisconsin*, 14: 79–80. Madison, Wis.: Democratic Printing Company, 1898.

*Unpublished Reports, Papers, and Theses*

Cooler, Kathleen Hudson E. "Hop Agriculture in Oregon: The First Century." Master's thesis. Portland State University, 1986.

Good, Sharon. "The Hop Culture." Master's thesis, State University of New York at Oneonta, 1968.

Hansen, Margaret Elizabeth. "The Hop Industry." Master's thesis, University of Washington, 1963.

Hibberd, Benjamin Horace. "The History of Agriculture in Dane County Wisconsin." Bulletin of the University of Wisconsin, no. 101. Ph.D. diss., University of Wisconsin, Madison, 1902.

Kuhler, Joyce Benjamin. "A History of Agriculture in the Yakima Valley, Washington, from 1880 to 1900." Master's thesis, University of Washington, 1940.

Mohney, Kirk Franklin. "Specialized Agricultural Architecture: Nineteenth Century Hop Houses in Oneida County, New York." Master's thesis, Cornell University, 1983.

Moraign, Evin C. "After the Riot at Wheatland." Sacramento State College, 1959. Typescript.

Parsons, James Jerome, Jr. "The California Hop Industry: Its Eighty Years of Development and Expansion." Master's thesis, University of California at Berkeley, 1939.

Sekora, Lynda. "Willamette Valley Hop Houses." University of Oregon, 1985. Typescript.

Tomlan, Michael A. "Hop Houses in Central New York, with Guidelines for Their Identification and Evaluation." Report prepared for the Madison County Historical Society and the New York State Office of Parks, Recreation and Historic Preservation, 1983. Photocopy.

## Manuscript Collections

Locke, William P. Papers. Archives, Colgate University, Hamilton, N.Y.

Meeker, Ezra. Papers. Box 3, Folder 23, Telegrams. Washington State Historical Society, Tacoma.

Tooke Family Papers. Department of Manuscripts and Archives, John M. Olin Library, Cornell University, Ithaca, N.Y.

White, Canvass. Papers, Department of Manuscripts and Archives, John M. Olin Library, Cornell University, Ithaca, N.Y.

# INDEX

England: influence on U.S. hop culture, 4–5, 189, 225
    (n. 5); problems with crops in, 25, 30, 103; London,
    36, 93, 95, 161, 217, 241 (n. 2); Kent, 55, 159, *160*,
    163, 164, *165*, *166*, 167, 172, 217; Farnham, 65;
    presses in, 73, 76; method of drying in, 162, *162*,
    163, *163;* Sussex, 164; introduction of hop culture to,
    226 (n. 1); market for U.S. product, 238 (n. 10), 239
    (n. 36). *See also* Great Britain
English Cluster hop, 50
E.O.L., 185, *186*
Erie Canal, 245 (n. 33)
Eugene, Oreg., 30, 32, *68*, 130
*Eugene Daily Guard*, 248 (n. 70)
Eugene Foundry & Machine Co., *68*
Europe, 6, 11, 13, 38–39, 73, 95, 103, 114, 161, 235
    (n. 60)
Exclusion Act (1882), 128, 129
Exportation, 12, 13, 14, 33, 38–39, 41, 62, 88, 95, 103
Ezra Meeker and Company, 78, 91, *93. See also*
    Meeker, Ezra

Fairfield, Wis., 25
Fan-blast kilns. *See under* Kilns
Farm Club (Wheatland, Calif.), 128
Farnham, England, 65
Farnsworth, Benjamin, 16
Fassett, Francis H., 208, *209*
Fayetteville, N.Y., 55
Federal Hop Control Board, 106
Federal legislation, 38, 39, 105, 106
Fernandez, Joseph N., 38
Ferris, Warren, 234 (n. 20)
Fertilizer, 45, 48, 50, 51, 85, 86, 87, 88. *See also* Soil
    conditions
Fire, danger of, *75*, 112, 175, 184, 192, 195, 196, 197,
    208–9, 222, 240 (n. 56)
Fireplaces, 161, 162, 164
Fitchen, John, 5
Flint, Daniel, 26–27, 34, 56, 127, 129, 194, 195, 230
    (n. 60), 237 (n. 73)
Flint, Wilson G., 26–27, 34, 230 (n. 60)
Flint and Raymond, 28, 196
Food Control Bill, 38
Ford, "Blackie," 152–54
Foreign pickers. *See* Pickers: city
Forest Grove, Oreg., 32
Fox & Searles, *93*
France, 13, 27, 233 (n. 18), 238 (n. 10)
France, Edward, 192–93, *193*
Franklin Co., N.Y., 19
Fruit driers, 39, 247 (n. 61)
Fuggle hop, 50, *50*
"Full-Armored" Royal sprayer, *61*

Fuller, Andrew Samuel, 183
Furnaces, 66, *68*, 69, *69*, 70, 159, 160, *160*, 167, 174,
    197, 202; in malt kiln, 163

Gates, LeRoy, 124
Genesee Country Village (Mumford, N.Y.), 219
*Genesee Farmer*, 24, 169–70
Germany, 13, 25, 38, 238 (n. 10)
Glacken and Wagner, 28, 196
Glassie, Henry, 4
Goshen, Oreg., 130
Gospel Swamp, Calif., 128
Gosse, Guy E., 101
Grafton Co., N.H., 14
Grande Ronde Reservation, Oreg., 130
"The Granger" hop stove, *68*
Grant's Pass, Oreg., 32
Grape hop, 50
Great Britain, 50, 65, 73, 159, 161, 162, 163, 167, 171,
    172, 217, 247 (n. 53). *See also* England
Green Valley, 27
Grouter, Louis: hop yard, *143*
Growers' costs, 36, 53, 69–70, 85, 86, 87, 88, 97, 104,
    109, 113, 114, 124, 126, 127, 128, 164, 180, 188. *See
    also* Pickers' pay
Growers' organizations, 100–105, *101*
Growing. *See* Hop growing
Guilford, N.Y., 54, 55
Gurley's Ranch, Wash., *58–59*

Haidas, 132
Hamilton, N.Y., 55, 85, 182, *183*
Hansen, G. J., 91
Hanson, C. J., 236 (n. 72)
Haraszthy, Count Augustin, 23
Harrington, H. Niles, 113
Harris, L. W., 77
Harris press, *76*, 77, 236 (n. 72)
Hartland, Wis., 191, *191*
Harvesting, 45, 49, 60, 85, 88; picking the hops, 50, 62,
    *63*, 64–66, 119–21, *120*, *122–23*, *125*, *135*, *136–37*,
    142–43, *143;* removing the poles, 51, 54–55, *63*, 64,
    *72*, 142–43; pole puller (box tender), 64, 85, 86, 87,
    *120*, 121, 142–43, 145; wire men, 64, 143, *143. See
    also* Mechanical pickers; Containers, pickers';
    Pickers: working conditions; Poles
Hawks (manufacturer), 80
Hayden and Lincoln, 32
Haynie, William M., 195–96
*Hearth and Home*, 221
Helmhold, H. C., 34
Henderson, J. C., 154
Henn, Minnie, 124

Marshall, N.Y., 126
Maryland: Baltimore, 17, 167
Marysville, Calif., 127, 242 (n. 23)
Massachusetts: beer production, 11–12; Middlesex
    Co., 12, *12*, 17, 160; Wilmington, *12*, 12–13, 168,
    *169*; Tewkesbury, *12*, 13; Townsend, *12*, 13; Boston,
    12, 13, 16, 89, 238 (n. 14); hop production, 12–13, 14,
    16, 19, *19*, *24*, 217, 227 (n. 20); Shirley, 13; hops
    legislation in, 13, 18; Stow, 17; quality of hops in, 18;
    screw presses in 73; use of charcoal-fired kiln in,
    161; early hop kilns in, 167–69, *168*
*Massachusetts Agricultural Journal*, 244 (n. 8)
Massachusetts Bay, 11
Maynard, Amos, 228 (n. 27)
Mead, E. C., 34. *See also* Thompson and Mead
Mechanical pickers. *See* Pickers, mechanical
Meeker, Ezra, 34, 35, 36, 56, 65, 87, 103, 198, 218, 228
    (n. 25), 231 (n. 89), 234 (n. 34); cooling barns and
    kilns, 72–73, *74–75*, 200, *201*, 202; hop field and
    camp, *134*
Meeker, Ezra, and Company, 78, 91, *93*
Meeker, Jacob R., 34, *37*
Meeker, M. J., 66
Meeker family, 34, 36, 88, 198. *See also* Puyallup Hop
    Company
Meis, Phillip, and Company, 95
Mendocino Co., Calif., 28–29, 30, 41, 62, 98, 127, *198*
Mendocino Hop Growers Association, 101
Mendocino-Sonoma hop district. *See* Mendocino Co.,
    Calif.; Sonoma Co., Calif.
Menke, George, 28, 196
Merrimack Co., N.H., 14
Michigan, 6
Middlesex Canal (Mass.), 13
Middlesex Co., Mass., 12, *12*, 17, 160
Midwest (midwestern U.S.), 23, 55, 178, 218
Milford, N.H., 13
Milwaukee, 103, 121, 124, 238 (n. 13)
Milwaukee and St. Paul Railroad, 121
Mississippi Valley, 91, 110
Moakler, John: hop house, 175, *178*
Mohawk Valley, 5
Monmouth Co., N.J., 185, *187*
Monroe, Wis., 121
Monroe Co., Wis., *195*
Montgomery Co., N.Y., 22
Moore Dry Kiln Company, 206, *208*
More, F., *53*
Morris, N.Y., 55, 183, *184*
Morrisville, N.Y., 170
Morrow, J. E., 246 (n. 34)
Morse, Albert W., 186, *187*
Moxee City, Wash., 36, 208, *209*
Moxee Company, 135

Mueller, C. G., 124
Mumford, N.Y., 219
Munnsville, N.Y., 171, *171*
Myers, Gardner, 25

Napa Co., Calif., 28
Nashua-to-Keene (N.H.) stage road, 13
Native American pickers. *See* Pickers: native
    American
Native hops, 5–6, 11, 23, 226 (n. 7)
Neame farm (Kent, England), 245 (n. 32)
Nevada, 129, 146
New Buffalo, Wis., 240 (n. 78)
*New Cyclopaedia; or, Universal Dictionary of Arts,
    Sciences, and Literature, The*, 161, 162, *163*
New England, 6, 11, 12, 13, 16, 18, 23, 24, *24*, 50, 62,
    66, *67*, 100, 106, 108, 119, 159, 172, 217, 228 (n. 38),
    229 (n. 53)
*New England Farmer*, 168, 170
New Hampshire, 13, 14, *14*, 16, 19, *19*, *24*, 26, 170
New Jersey, 5, 185, *187*
Newport, N.H., 170
Newport, Wis., 113
Newton, W. S., 237 (n. 81)
New York (State): Cooperstown, v, 19, *54*, 90, 107,
    108, 109, *109*, 110, *123*, 126, 145, 175, *178*, *181*, 219;
    material culture of, 4; barns in Mohawk Valley, 5;
    hop production, 6, 18–19, *19*, 22–23, *24*, 26, *28*, 29,
    *33*, 85, 86, 87; New York City, 16, 17, 18, 26, 28, 80,
    93, 95, 97, 221, 238 (nn. 10, 26); Brooklyn, 17;
    Bouckville, 17, *17*, 107, 172, *173*; Madison Co., *17*,
    17–18, 19, *68*, 86, 121, 124, 170, 171, *171*, 182, *183*,
    186, *187*, 241 (n. 8), 245 (nn. 33, 34); Madison, *17*,
    113, 219, 228 (n. 27); Oneida Co., N.Y., 18, 19, *72*,
    90, 121, 144, *144*, 172, *173*, 174, 175, *176*, 190, 209,
    *210*, 233 (n. 14), 240 (n. 56), 241 (n. 8), 245 (n. 26),
    246 (n. 34); Otsego Co., 18, 19, 107, *108*, 113, 120,
    121, 124, 143, 174, 175, 178, *178*, 181, *181*, 183, *184*,
    228 (n. 35), 236 (n. 38), 241 (n. 8); hops legislation in,
    18, 65, 105; Albany, 18, 108; Sangerfield, 18, 126,
    240 (n. 56); Franklin Co., 19; Ontario, 19;
    Waterville, 19, *20–21*, 21, 22, *22*, 24, *64*, *68*, *76*, 77,
    78, *80*, 80–81, 87, 90, *90*, 93, 100, *101*, 112, 113, 126,
    145, 170, 175, *176*, 235 (n. 60), 236 (n. 72); Herkimer
    Co., 19, 121, 241 (n. 8); Schoharie Co., 19, *193*;
    Montgomery Co. 22; Syracuse, 27, 121, 126; Clinton
    Co., 36; Constable, 36; during Prohibition, 39; hop
    varieties, 50; Adirondack Mountains, 53; Guilford,
    54, 55; Barrington, 55; Fayetteville, 55; Pitcher, 55;
    Sharon, 55; Starkville, 55; Warnersville, 55; Yates
    Co., 55; horizontal hop growing, 55, 56; Hamilton,
    55, 85, 182, *183*; Morris, 55, 183, *184*; Cobleskill, 55,
    193, *193*; hop mildew, 60; hop harvest, 62; hop
    boxes, 65; early kilns, 66, 235 (n. 51); Leonardsville,

New York (State) (*cont.*)

68; cost of drying in, 69–70; Long Island City, 80; expenses of growers, 85–88; East Hamilton, 86; Jefferson Co., 90; Schoharie, 90; Norwich, 100; Sylvan Beach, 100; growers' cooperative, 105; scale of operations, 106; Oswego, 107; Hyde Park, 108; Toddsville, 108; Phoenix, 108, *108*, 219; Hop City, *108*, 108–9; Utica, 110, 121; Peterboro, 113; Eaton, 113, 186, *187*, 228 (n. 27); home pickers, 119; Cazenovia, *120;* Oneida, 121; Oswego Co., 121; West Monroe, 121; Central Square, 124; hop structures, 169–85; Morrisville, 170; Munnsville, 171, *171;* Bridgewater, 172, *173*, 209, *210;* Augusta, 174, 228 (n. 27), 233 (n. 14), 245 (n. 26); Improved Hop-Dryer, 192–93; evolution of construction in, 214; development of hop culture in, 217; Mumford, 219; Lake Ridge, 221; Tompkins Co., 221; Orleans Co., 229 (n. 48); hop prices, 229 (n. 53); hops bleaching, 235 (n. 60); Pratts Hollow, 245 (n. 27); Sullivan, 245 (n. 33); Jamestown, 248 (n. 73). *See also* East

New York, Ontario and Western Railroad, 121

New York and Oswego Railroad, 121

New York City, 16, 17, 18, 26, 28, 80, 93, 95, 97, 221, 238 (nn. 10, 26)

New York Extract Company, *80*, 80–81

*New York Herald*, 221

*New York Observer*, 221

*New York Tribune*, 221

Noble, Allen, 4–5

North America, 4, 6, 11, 12

Northeast (northeastern U.S.), 4, 6, 17, 66, 119

*Northern Farmer*, 170

Northern Pacific Railroad, 34, 36, 103, 135

North Puyallup, Wash., 135

Northwest (northwestern U.S.), 30, 36, 37, 53, 57, *58*, 66, 132, *135*, 234 (n. 34)

Northwest Brogdex Company, 208, *209*

North Yakima, Wash., 110

Norwich, N.Y., 100

Oakland, Calif., 229 (n. 55)

Oasts and oast houses, 159, 160, *160*, 161–62, 163, 164, *165*, *166*, 167, 172, *173*, 174, 186, *187*, *189*, 190, *190*, 217, 225 (n. 5). *See also* Kilns

Off pickers. *See* Pickers: city

Olympia, Wash., 34

Oneida, N.Y., 121

Oneida Chief Farms, 209, *210*

Oneida Co., N.Y., 18, 19, *72*, 90, 121, 144, *144*, 172, *173*, 174, 175, *176*, 190, 209, *210*, 233 (n. 14), 240 (n. 56), 241 (n. 8), 245 (n. 26), 246 (n. 34)

Ontario Co., N.Y., 19

Oregon: hop production, 6, *28*, 30, *30*, 32, 33, *33*, 34,

39–40; suitability for hop growing, 26; Eugene, 30, 32, *68*, 130; Albany, *30;* Willamette, *31;* Salem, *30*, 32, 98, 104, 130, 131, 208, 238 (n. 5); Buena Vista, 30, 238 (n. 5); Benton Co., 32; Clackamas Co., 32; Forest Grove, 32; Grant's Pass, 32; Linn Co., 32; Washington Co., 32; Marion Co., 32, 33, 96; Lane Co., 32, 33, 102; Polk Co., 32, 33, 202; Springfield, 32, *46–47*, 130, 202, *205;* Portland, 32, 95, *99*, 130–31, 154, 208, 239 (n. 41); Oregon City, 32, 203; Corvallis, 32, 203, 209, *213;* Independence, 32–33, 57, *79*, 102, 110, 139, *139*, 154, 206, 208, 231 (n. 87), 233 (n. 11); "Bird Island," 33; competition with Washington, 34; acreage devoted to hop culture, 39, 41; use of short poles, 56; hop mildew, 60; hop harvest, 62; hop containers, 65–66; presses, 77; Santa Clara, *77;* Silverton, 98; growers' organizations, 101–3; Woodburn, 102, 235 (n. 53); hop marketing, *105*, 106; recruiting pickers in 111; Creswell, 130; Goshen, 130; Grande Ronde Reservation, 130; Warm Springs Reservation, 130; Astoria, 131; working conditions, 154–55; Bottom, 197; names for hop houses, 199; Santiam Crow, *206;* frame kilns, 208; twin fan installation, *208;* activities of Richard Weaver, 238 (n. 13). *See also* Northwest; Pacific Coast; Pacific Northwest; West

Oregon-California-Washington Hop Marketing Agreement, *105*, 105–6

Oregon City, Oreg., 32, 203

Oregon Hop Growers' Association (1899), *99*, 102–3

Oregon Hop Growers Association (1932), 104

Oregon State Agricultural Society, 32

Oregon State College, Corvallis: Agricultural Experiment Station, 209, *213*

Oregon State Fair, 30, 32

Oregon Trail, 218

Orleans Co., N.Y., 229 (n. 48)

Orleans Co., Vt., 14

Osborne, Charles, 113

Oswego, N.Y., 107

Oswego Co., N.Y., 121

Otsego Co., N.Y., 18, 19, 107, *108*, 113, 120, 121, 124, 143, 174, 175, 178, *178*, 181, *181*, 183, *184*, 228 (n. 35), 236 (n. 68), 241 (n. 8)

Oxford Co., Me., 14, *15*

Pacific Coast, 26, *28*, *33*, 34, 39, *55*, 64, 70, 81, 103, 104, *105*, 105–6, 110, 111, 115, 126, 128, 144, 195, 196, *210*, 214, 219, 247 (n. 61)

Pacific Co., Wash., 36

Pacific Northwest, 6, 30, *37*, 65, *75*, 77, 91, 95, 197, 202, 209, 218

*Pacific Rural Press*, 196

"Paddock Machine" (hop picker), 113